Banking on Milk

Banking on Milk takes the reader on a journey through the everyday life of donor human milk banking across the United Kingdom (UK) and beyond, asking questions such as the following: Why do people decide to donate? How do parents of recipients hear about human milk? How does milk donation impact on lifestyle choices?

Chapters record the practical everyday reality of work in a milk bank by drawing on extensive ethnographic observations and sensitive interview data from donors, mothers of recipients and the staff of four different milk banks from across the UK, and visits to milk banks across Europe and North America. It discusses the ongoing pressures to do with supply, demand and distribution. An empirically informed "ethnography of the contemporary", where both biosociality and biopower abound, this book includes an exploration of how milk banks evolved from registering wet nurses with hospitals, showing how a regulatory culture of medical authority began to quantify and organize human milk as a commodity.

This book is a valuable read for all those with an interest in breastfeeding or organ and tissue donation from a range of fields, including midwifery, sociology, anthropology, geography, cultural studies and public health.

Tanya Cassidy is a Fulbright-HRB (Irish Health Research Board) Health Impact Scholar, an EU Horizon 2020 Marie Skłodowska Curie Award (MSCA) fellow and an Irish Health Research Board Cochrane Fellow. She is a Visiting Fellow at the University of Central Lancashire (UCLan) where she held her MSCA.

Fiona Dykes is Professor of Maternal and Infant Health and leads the Maternal and Infant Nutrition and Nurture Unit (MAINN), University of Central Lancashire. She is an Adjunct Professor at Western Sydney University and holds Visiting Professorships at Högskolan, Dalarna, Sweden and Chinese University of Hong Kong.

Bernard Mahon is Professor of Immunology and Cell Biology at the National University of Ireland Maynooth (Maynooth University). Currently, he is collaborating with a European consortium examining immunological crosstalk between mother and neonate, and scientific aspects of human milk exchange. He is a former chairperson of the Maynooth University Research Ethics Review Board.

Routledge Studies in the Sociology of Health and Illness

Recovery, Mental Health and Inequality
Chinese Ethnic Minorities as Mental Health Service Users
Lynn Tang

Fertility, Health and Lone Parenting
European Contexts
Edited by Fabienne Portier-Le Cocq

Transnationalising Reproduction
Third Party Conception in a Globalised World
Edited by Róisín Ryan Flood and Jenny Gunnarsson Payne

Public Health, Personal Health and Pills
Drug Entanglements and Pharmaceuticalised Governance
Kevin Dew

Dementia as Social Experience
Valuing Life and Care
Edited by Gaynor Macdonald and Jane Mears

Injecting Bodies in More-than-Human Worlds
Mediating Drug-Body-World Relations
Fay Dennis

Contested Illness in Context
An Interdisciplinary Study in Disease Definition
Harry Quinn Schone

Banking on Milk
An Ethnography of Donor Human Milk Relations
Tanya Cassidy and Fiona Dykes with Bernard Mahon

For more information about this series, please visit: www.routledge.com/Routledge-Studies-in-the-Sociology-of-Health-and-Illness/book-series/RSSHI

Banking on Milk

An Ethnography of Donor
Human Milk Relations

**Tanya Cassidy and Fiona Dykes,
with Bernard Mahon**

LONDON AND NEW YORK

First published 2019
by Routledge
2 Park Square, Milton Park, Abingdon, Oxon OX14 4RN

and by Routledge
52 Vanderbilt Avenue, New York, NY 10017

Routledge is an imprint of the Taylor & Francis Group, an informa business

© 2019 Tanya Cassidy and Fiona Dykes, with Bernard Mahon

The right of Tanya Cassidy and Fiona Dykes to be identified as authors of this work has been asserted by them in accordance with sections 77 and 78 of the Copyright, Designs and Patents Act 1988.

All rights reserved. No part of this book may be reprinted or reproduced or utilised in any form or by any electronic, mechanical, or other means, now known or hereafter invented, including photocopying and recording, or in any information storage or retrieval system, without permission in writing from the publishers.

Trademark notice: Product or corporate names may be trademarks or registered trademarks, and are used only for identification and explanation without intent to infringe.

British Library Cataloguing-in-Publication Data
A catalogue record for this book is available from the British Library

Library of Congress Cataloging-in-Publication Data
A catalog record for this book has been requested

ISBN: 978-1-138-55907-3 (hbk)
ISBN: 978-0-203-71305-1 (ebk)

Typeset in Times New Roman
by Apex CoVantage, LLC

Contents

Foreword RIMA D. APPLE		vi
About the authors		ix
Acknowledgements		x
Abbreviations		xii
Preface		xiv
	Introduction	1
1	Ethnography of human milk exchange in the contemporary world	2
2	Moving hospital wet nurses to bureaus and banks	23
3	Building the science and society of human milk with banks	55
4	"It's Not Rocket Science": practice and policy in human milk banking	73
5	Pumping for preemies	88
6	Building liquid bridges	101
	Endword TANYA CASSIDY	116
	Index	124

Foreword

"Jamie, the poor little fellow is six months old and has been sick almost since birth", wrote Julia Carpenter in her diary in January 1889. Living on the Dakota Territory, US, in the late nineteenth century, and unable to nurse her child, she discovered many suggestions for substitutes for mother's milk. Over the months, she fed Jamie a variety of recipes for modified cow's milk formulas and commercial products, but to no avail. She travelled to Aberdeen where she was able to hire a wet nurse who nursed the infant during the day; at night, he was fed condensed milk. After two months, she needed to return home, where Jamie received a series of patent medicines and even wine. One March morning, Jamie awakened his mother at 6 am. Carpenter reported:

> He seemed to breathe funny yet I warmed his dinner and gave him but he did not take it although I had thought he took the wine. He kept gasping.... I took him in my arms calling Jamie, Jamie, he did not look at me, but his eye gradually closed, he gasped a few times, and was *dead, my* baby, *my* darling, *my* boy Jamie all without a moments warning.

James Lucien Carpenter died at age eight months.

Stories such as this from diaries and letters document the experiences of women who have long needed and sought a substitute for mother's milk. Mrs. H.S.K. beseeched the popular women's magazine the *Ladies' Home Journal* in 1901 for advice on feeding her two-week-old infant whom she had been forced to stop nursing. In a 1907 issue of *American Motherhood*, A.W. bemoaned her situation:

> When my first baby came, I had plenty of good milk for four weeks, then the quality became poor and the quantity diminished until at six weeks I had to wean him entirely.

Others used the columns of women's magazines to exchange histories and counsel for worried mothers searching for an alternative method of infant feeding.

Because their infants were premature or sickly or they felt their own milk supplies were insufficient, some mothers have needed assistance in nourishing their children. In the past, women resorted to different animal milks augmented with gruels. Archeological evidence of infant feeders documents the use of such alternatives back to at least the second century CE. By the nineteenth century, commercial infant foods had begun to appear in the market. These many options—the increasing availability of commercially manufactured formulas and widely circulated recipes for home-prepared foods and those designed by physicians—were intended as alternatives to mother's milk, which, to quote one mother in the popular magazine *Babyhood* in 1886, was the "natural and proper food" for an infant. By the twentieth century, the overwhelming employment of bottle feeding in Western societies did not completely negate the desire and need for breastfeeding infants who lacked mother's milk. Some hospitals employed wet nurses; historically, in some societies, a relative or friend who was lactating or a wet nurse, often a woman paid for this service, supplied this nutrient.[1]

The situation is somewhat different today. We still consider breastmilk as the best food for infants: advice proclaimed by the World Health Organization, UNICEF and numerous other international, national and professional organizations. Though bottle formulas continue to be widely utilized around the globe today, many women who cannot provide mother's milk for their child reject the artificial. They instead find the answer to this dilemma in the breastmilk of another mother, which is increasingly delivered by a human milk bank.

Human milk banks: to some an unfamiliar concept, to others a matter of life or death. Though these institutions are not a new phenomenon, they have multiplied dramatically in this century. In principle, the concept of the human milk bank is simple. Lactating women contribute their breastmilk to nurture infants deprived of this sustenance. Yet, their shape, their procedures, their operations vary widely across the world, affected by local needs, local governmental and healthcare structures and local culture. Many questions surround the pragmatics of a human milk bank. First and foremost is the source of the milk. The banks in the UK that Tanya Cassidy and Fiona Dykes study depend on contributions from volunteers, women who often personally experienced or observed the benefits of human milk banks and who typically view their donations as gifts. In other countries, there are banks that pay women for their milk. Then there is the concern about the safety of the milk. Who establishes the standards of collection, transporting and preservation of the milk? Moreover, there is no single criterion for any of these processes. Should the milks of multiple women be pooled, or should each contributor's milk be kept uniquely identified, a practice sometimes determined by local culture? Some banks pasteurize the milk, but others do not. What is the best method for preserving the milk? Where should banks be located? In hospitals? In community clinics? Are human milk banks at base clinical entities or research laboratories? In some countries, governmental agencies have oversight for this formal exchange of breastmilk; in others, women buy and sell breastmilk in an informal market with fewer protections.

As a detailed ethnographic study of the four largest donor human milk banks in the UK, *Banking on Milk* enables us to appreciate the complexity of any system that seeks to distribute human milk to needy infants. There is no one-size-fits-all for successful banks; the institutions must develop within the conditions that necessitate their founding. Once established, they are not fixed but must continue to evolve as their situations change. The diversity of banks studied confirm the difficulty of assigning any simple description for these institutions. But this lack of a single definition should not discourage us from creating human milk banks, but rather it should challenge us, as Cassidy and Dykes document in this important book, to design human milk banks responsive to their communities.

Rima D. Apple, Ph.D.
Professor Emerita
University of Wisconsin–Madison

Note

1 For more on this history, see Rima D. Apple, *Mothers & medicine: A social history of infant feeding, 1890–1950* (Madison: University of Wisconsin Press, 1987).

About the authors

Tanya Cassidy is a Fulbright-HRB (Irish Health Research Board) Health Impact Scholar, an EU Horizon 2020 Marie Skłodowska Curie Award (MSCA) fellow and an Irish Health Research Board Cochrane Fellow. She is a Visiting Fellow at the University of Central Lancashire (UCLan) where she held her MSCA. Tanya recently took up a lectureship in the School of Nursing and Human Sciences at Dublin City University (DCU) in Ireland and continues to work closely with her colleagues in the Department of Anthropology at Maynooth University.

Fiona Dykes is Professor of Maternal and Infant Health and leads the MAINN School of Community Health and Midwifery, University of Central Lancashire, which she established in 2000. She is also an Adjunct Professor at Western Sydney University and holds Visiting Professorships at Högskolan, Dalarna, Sweden and Chinese University of Hong Kong. Fiona has a particular interest in the global, socio-cultural and political influences on infant and young child feeding practices.

Contributions to Chapter 3 by Bernard Mahon, Professor of Immunology and Cell Biology at the National University of Ireland Maynooth (Maynooth University). His research focuses on aspects of neonatal immunity to infection, immunization and new cell therapy. He has worked on preclinical and clinical trials for novel neonatal whooping cough vaccines. Currently, he is collaborating with a European consortium examining immunological crosstalk between mother and neonate, and scientific aspects of human milk exchange. He is a former chairperson of the Maynooth University Research Ethics Review Board.

Acknowledgements

All ethnographic studies are, by definition, collaborations, and we are tremendously indebted to a large number of people, many of whom we cannot thank by name, as they were participants in our data collection and therefore for ethical reasons must remain anonymous. Included in this unnamed debt of gratitude are all the mothers who contributed to our ethnographic research project by giving their time so generously and telling us their stories about donor human milk either as donors or as parents of recipients, as well as staff members whose expertise by experience has proved key to our ethnographic knowledge construction. Although many of the staff in the donor human milk services involved in our research also gave us interviews, we would like to thank some people by name. Specifically, we wish to thank, from the Western Trust Human Milk Bank in Northern Ireland, Ann McCrea, Vivienne Carson and Wendy Oldfield, as well as several other staff affiliated with the South West Acute Hospital in Enniskillen. We are also extremely grateful to the team at the One Bank for Scotland service and would like to extend our thanks to Debbie Barnett, Jess MacFarlane, Grace Brown and Judith Simpson, as well as to members of the staff and especially the important infant feeding team at the National Health Service Greater Glasgow and Clyde Queen Elizabeth University Hospital (where this donor human milk service is located), including Joyce Young, Gillian Bowker and Linda Wolfson, who is now the professional advisor and national maternal and infant nutrition co-ordinator for the Scottish government. In addition, we wish to thank the team at the wonderful Northwest Human Milk Bank located at the University of Chester, including Jackie Hughes, Emma Savage, Stacey Griffiths, Gillian Brady and Carol Barnes, as well as a number of members of staff at the Countess of Chester Hospital who also helped with our data collection. The fourth donor human milk service involved in our research, and our lead research and development team, is part of the Imperial College Health NHS Trust, and is located next to the neonatal intensive care unit (NICU) at Queen Charlotte and Chelsea Hospital (QCCH). We wish to thank Gillian Weaver, who has been a support throughout the development of this work and who retired after almost three decades of managing the QCCH service around the time we were about to begin conducting our ethnographic data collection, but who nonetheless continued to support our research. We also wish to thank Dr Natalie Shenker, Dr Stephanie D'Arc, Ruta Janaviciute and Laila Chouita, as well as other members of staff at QCCH and St Mary's who helped us during our research collection, including obtaining ethical approval. In addition to all the people involved at these four donor human milk services, we also wish to thank everyone involved in the UK Association of Milk Banking (UKAMB). Some of these inspiring individuals also took the time to tell us their stories about donor human milk and, therefore, again for ethical reasons, we will not name them specifically.

Over the years, we have worked very closely with a number of members of the European Milk Banking Association (EMBA) and would like to extend our special gratitude to Aleksandra Wesolowska in Poland, Rachel Buffin and Jean Charles Picaud in Lyon, who all contributed to our EU MSCA research project. We would also like to extend our gratitude to Clair Yves Boquien (France), Anne Grøvslien (Norway), Antoni Gaya (Spain), Corinna Gebauer (Germany), Sertac Arslanoglu (Turkey) and Guido Moro (Italy), as well as the current president of EMBA, Enrico Bertino (Italy), whose work on comparative guidelines was a catalyst for our original research idea.

We have also been very fortunate to interact with people involved in the international milk banking world over the last decade, and we would also like to thank a number of people within the international milk banking community, in particular Penny Reimer and Anna Coutsoudis from South Africa, Ben Hartmann from Perth, Australia, as well as Kiersten Israel-Ballard and Kimberly Mansen (Amundson) from PATH. We also wish to extend our sincere gratitude to all the other individuals in the international world of donor human milks services with whom Tanya in particular has met and interacted with at conferences. We hope we will continue to collaborate with them in the context of future research and writing projects. We consider ourselves to be very fortunate to have interacted with so many of you over the years and appreciate that your varied knowledge of in the global world of donor human milk services underlies our understanding and this ethnographic discussion.

We would also like to thank all of our colleagues at the University of Central Lancashire, especially Gill Thomson whose help has been invaluable throughout this project. Also all of the other members of the MAINN team, including the wonderful students, whose collegial interaction was invaluable. Tanya would also like to thank her colleagues in the Department of Anthropology at Maynooth University and all her other colleagues at MU, as well as her new colleagues in the School of Nursing and Human Sciences at Dublin City University.

As part of our research, we collected archival materials, and we also wish to thank the archivists at the Wellcome Trust Library, the British Library, Queen Charlotte and Chelsea Hospital and Graham Deacon at Historic England.

Please note that the research which underpins this book was enabled, directly and indirectly by funding from the EU Horizon 2020 Marie Skłodowska Curie Award (654495), as well as benefiting from the generous patronage of Drenda Vijuk in the loving memory of Joseph Vijuk.

Finally, we would also like to acknowledge the support provided by all our own donor human milk relations, including our own mothers, fathers, partners and children, as well as all the significant others in our lives, creating the dynamic protean familial support networks without which no intellectual enterprise is possible, or indeed, worthwhile.

Abbreviations

AIDS	Acquired Immune Deficiency Syndrome
ALCI	Association of Lactation Consultants of Ireland
BAPM	British Association of Perinatal Medicine
BMJ	British Medical Journal
CIC	community interest company
DCU	Dublin City University
DHM	donor human milk
EMBA	European Milk Banking Association
EU	European Union
HIV	Human Immunodeficiency Virus
HMBANA	Human Milk Banking Association of North America
HMO	Human Milk Oligosaccharides
HoP	Holder Pasteurization
HRA	UK Health Research Authority
HRB	Irish Health Research Board
HTLV	Human T-cell lymphotropic virus
HTST	High-temperature Short-time
IBCLC	International Board Certified Lactation Consultant
ICCBBA	International Council for Commonality in Blood Banking Automation
ISBT	International Society of Blood Transfusion
LLL	La Leche League
MOM	mother's own milk
MSCA	EU Horizon 2020 Marie Skłodowska-Curie Award
MU	Maynooth University
MUIMME	*M*ilk Banking and the *U*ncertain *I*nteraction between *M*aternal *M*ilk and *E*thanol
NEC	necrotizing enterocolitis
NHS	National Health Service
NICE	National Institute for Health and Care Excellence
NMB	National Milk Bank
PATH	Program for Appropriate Technology in Health
QCCH	Queen Charlotte and Chelsea Hospital
SfAA	Society for Applied Anthropology

UCLan	University of Central Lancashire
UK	United Kingdom
UNICEF	United Nations International Children's Emergency Fund
UKAMB	UK Association for Milk Banking
WBTi	World Breastfeeding Trends Initiatives
WHO	World Health Organization

Preface

Marilyn Strathern visited Maynooth University (MU) as we were finalizing this book, and it was the first time I was able to meet her. Her classic discussion about gender and gift(s) in Melanisia (Strathern 1988, 2016) has always framed my reading of Marcel Mauss' classical anthropological framing of gift exchanges (1925, 2016). This coincidence adds to the additional publication coincidence we discuss more in the next chapter of Strathern's volume titled *Before and After Gender* and an expanded translation of Mauss' discussion of the gift. In her Maynooth talk, Strathern (2019) reminds us that in Melanesian performances, it is the performer who is expected to pay his audience if he causes them to feel emotions, in other words, if they perform well. It is this surprise of exchange which Strathern has encouraged her readers to seek and this part of what we hope our reader experiences.

Based on one of the largest multi-sited ethnographic studies of donor human milk services ever to be conducted by senior researchers, we frame our discussion not only ethnographically but also within considerations of global health research, particularly in light of the rapid expansion of donor human milk services around the world. Borrowing from Prentice (2010), we subscribe to the view that ethnographic research in global health research has four main principles:

1 It uses fieldwork to build theory.
2 It emphasizes meaning and classification.
3 It explores the negotiated nature of reality.
4 It emphasizes the central role of context.

As part of this view, we see ethnographic research as potentially challenging taken-for-granted views of the world and specifically look at our own place in those worlds. In a health research context, we are reminded that interventions are dynamic with unintended consequences that are socially constructed, involving power and negotiations, even contestations, between groups (Kleinman 2010). To the ethnographer, a health intervention or a service is not simple or straightforward. It is not neutral and neat; it is often very complex and even messy. Ethnographies are anchored in histories, politics and relations involving exchanges. *Banking on Milk* is a collaborative ethnographic narrative about the potential complexity of relations involved in the donor human milk exchange. Like all ethnographies, it is only one story, situated in time and space. It is indebted to all those who gave so generously of their time and knowledge, all of whom could, and perhaps some will, tell a potentially very different story about the relations underlying donor human milk exchange. We offer this narrative to help to expand the knowledge about the cultural worlds of donor human milk relations.

We begin by setting the scene by exploring the theoretical complexity of our ethnography of the "contemporary" world of donor human milk exchange, where the contemporary is envisioned as "a moving ratio of modernity, moving through the recent past and near future in a (non-linear) space" (Rabinow 2007, 2). Gifts have been more traditionally theorized in oppositional terms to commodification, but this chapter offers a more complex and synergistic understanding of gift and commodity relations. Our research will help to reconfigure understandings of maternal/corporeal generosity (Diprose 2002) in terms of a re-theorization of exchange. In this chapter, we also briefly discuss our interdisciplinary and international EU-funded MUIMME ethnographic project, and our personal collaboration regarding donor milk banking across the UK. Fiona's long-standing and international reputation in Maternal and Infant Nutrition and Nurture (MAINN) research offered transdisciplinary extensions for this research (Dykes 2006). We also offer a brief outline of each chapter.

In Chapter 2, we map the role of the hospital wet nurse, once widespread throughout the UK medical system, but highly criticized when associated with high rates of infant mortality, especially within so-called orphan or "foundling" hospitals. The gendering of professional roles in healthcare settings affects the conflictual expansion of infant feeding for "weaklings" (as prematurely born infants were once widely known in the medical world) whose celebrity was spreading globally. International differences associated with risk assessment and pasteurization change following wartime concerns resulted in the establishment of Human Milk Bureaus in the UK, while in North America, there was a move towards banking on bodies (Swanson 2014). Eventually, bureaus in the UK NHS system became known as "banks", which we describe using archival materials regarding the origins of the current four largest UK human milk banks.

In Chapter 3, we discuss non-linear links between contemporary breastmilk science and human milk banking. For over a century, research has explored the medicinal and preventative qualities of breastmilk with reference to intestinal infections, such as necrotizing enterocolitis. Also included in this discussion will be anti-viral and anti-bacterial properties of human milk, along with more recent scientific research that endeavours to identify cancer fighting properties in milk. In addition, researchers are exploring breastmilk as a potential source of stem cells. It should be noted that chemical analyses have not always been in the service of a pro-breastfeeding agenda and have in fact been sponsored by attempts to synthesize various milk formulas. To posit breastmilk as "naturally" suited to the demands of vulnerable infants has often meant subjecting breastmilk to the most scientific and clinical of interventions. In turn, these scientific constructions have recently been applauded by lay communities of breastfeeding groups, arguing that the scientific frame removes moral undertones, a point we critically interrogate.

In Chapter 4, we explore the everyday life of a donor human milk bank. Organized around the definition of a milk bank, we discuss collection, screening, processing, storing and distribution of donor human milk every day. The working regime, the practical environment and the staffing of milk banks create very specific social environments and interactions which normalize particular pressures and priorities. By considering "time" and "space", two highly theorized concepts in the social sciences and the humanities, we offer interpretations of their dearth within the worlds of human milk banking.

In Chapter 5, we explore how donor human milk banking involves establishing a community of generosity (Cassidy 2012). This chapter argues that mothers or recipient babies are not the passive beneficiaries of milk banking but have contracted into a network of relationships governed by a common sense of an urgent need for human milk for human babies. The motives, experiences and discoveries of donor and recipient mothers form the heart of this chapter.

In Chapter 6, we will discuss equity of access, which involves considerations of not only who receives milk but also in terms of who supplies the milk. We begin by discussing equity of access across the UK (which increasingly means across borders). Comparative discussion will involve European and global examples, making particular use of the special case of Brazil. As part of this equity of access, we need to think about where milk comes from and where it goes to, the nature of the exchanges between all women who are able to contribute and all children who are in need and should this situation remain in the control of a medicalized authority. International models of access will be part of our discussion. The central themes of trust translation and technology are all components of this more global perspective.

Finally, we conclude the book with an endword written by Tanya, which tells her personal journey into the world of donor human milk services. This is a collaborative ethnography, and we are both reflexive critical ethnographers, but only one of us has an affective link to milk banking, although we are both mothers. In the past, I have told my story at the beginning, including in my first publication related to milk banking and milk kinship, which has a narrative autoethnographic link between Ireland and the Sudan, and tells the story of two sons born 50 years apart, Gabriel and Mohamed (Cassidy and El Tom 2010). Later, I would apply autoethnographic tools and link this to literary poetics and publish with my husband, whose expertise is on literary poetry, but who also composed a poem entitled "The Milk Man" with which we began our narrative paper (Cassidy and Brunström 2015).

References

Cassidy, Tanya M. 2012. "Mothers, Milk, and Money: Maternal Corporeal Generosity, Sociological Social Psychological Trust, and Value in Human Milk Exchange." Special Issue on Motherhood and Economics. *Journal of the Motherhood Initiative (JMI)* 3 (1): 96–111.

Cassidy, Tanya and Abdullahi El Tom. 2010. "Comparing Sharing and Banking Milk: Issues of Gift Exchange and Community in the Sudan and Ireland." In *Giving Breast Milk: Body Ethics and Contemporary Breastfeeding Practice*, edited by Alison Bartlett and Rhonda Shaw, 110–21. Toronto: Demeter Press.

Cassidy, Tanya and Conrad Brunström. 2015. "Production, Process and Parenting: Meanings of Human Milk Donation." In *What's Cooking Mom? Narratives About Food and Family*, edited by Tanya M. Cassidy and Florence Pasche Guignard, 58–70. Toronto: Demeter Press.

Diprose, Rosalyn. 2002. *Corporeal Generosity: On Giving with Nietzsche, Merleau-Ponty, and Levinas*. Albany, NY: SUNY Press.

Dykes, Fiona. 2006. *Breastfeeding in Hospital: Mothers, Midwives and the Production Line*. London: Routledge.

Kleinman, A. 2010. "Four Social Theories for Global Health." *Lancet* 375 (9725): 1518–19. pmid:20440871

Mauss, Marcel. 1925. *Essai sur le don: forme et raison de l'échange dans les sociétés archaiques* l'Année Sociologique, seconde série, 1923–1924.

———. 2016. ed., annot. & trans. Jane I. Guyer, foreward Bill Maurer, Chicago: Hau Books.

Prentice, R. 2010. "Ethnographic Approaches to Health and Development Research: The Contributions of Anthropology." In *The SAGE Handbook of Qualitative Methods in Health Research*, edited by I. Bourgeault, R. Dingwall, and R. De Vries, 157–73. London: Sage Publications Ltd.

Paul, Rabinow. 2007. *Marking Time: On the Anthropology of the Contemporary*, Princeton, NJ: Princeton University Press.

Strathern, Marilyn. 1988. *The Gender of the Gift Problems with Women and Problems with Society in Melanesia*.

———. 2016. *Before and After Gender*, edited with an Introduction by Sarah Franklin, Afterword by Judith Butler. Chicago: HAU Books.

———. 2019. "Finding the Unlooked-for: Tricking Oneself and the Cultivation of Surprise." Talk Given at Maynooth University Ethnography Winter School.

Swanson, Kara W. 2014. *Banking on the Body: The Market in Blood, Milk, and Sperm in Modern America*. Cambridge, MA: Harvard University Press.

Introduction

Donor human milk banks are expanding around the world at an exponential rate, which is directly linked to global increases in premature births. The importance of human milk for prematurely born infants has been extensively identified, even among the recent social scientific work that has questioned the efficacy of human milk and health considerations. In addition, research also shows that a significant percentage of these mothers, at least initially, experience lactation problems. Europe is taking a leadership role in the expansion of human milk banks, although issues associated with alcohol consumption and maternal donations are a concern for clinicians and healthcare staff, given the increasing problems associated with drinking among women of childbearing age. The UK with its long history and current global leadership role is an ideal place to study these considerations, which will inform these larger issues of human milk for the prematurely born infant. The country is a leader in this century-old intervention, supporting not only one of the oldest hospital-based banks in Europe but also an important cross-border collaboration on the island of Ireland, along with a research-based national bank in Scotland, each representing different cases contributing significantly to the re-birth of the medical control of human milk. The UK is poised to offer the world vital information regarding donor human milk banking, maternal bodies and "trust". An important psycho-social theoretical concept is used to frame the triangulated data collected (including interviews, archival data and ethnographic information). In supporting an excellent experienced female researcher to return to the academy following a maternity/career break, this work directly supports women and science in society. Our EU Horizon 2020 project was called MUIMME, an old Irish word for wet nurse.

1 Ethnography of human milk exchange in the contemporary world

Certain research topics are chosen because of an urgency of societal need; others are chosen because of their intellectual fascination. Donor human milk services involve both, insofar as they are a response to a large-scale, life-saving intervention with massive public policy implications and at the same time a theoretically complex reflection on what it means to be a woman as well as a mother, thus provoking larger social and cultural questions regarding what it means to be a social being. Milk donation provokes philosophical debates about the limits of personhood and the extent to which something one produces is something one owns. Nor do these philosophical issues function in rarefied isolation from public policy debate. Central to the recruitment of milk donors is a public awareness campaign that confronts and addresses the so-called ick factor – a gut reaction to an unfamiliar reconfiguration of maternal responsibility which activates (and re-activates) alternative economic models of reciprocity and exchange.[1] The normalizing of human milk exchange practices in various parts of the world suggests that specific ideological formations are responsible for determining what is "instinctively" felt to be natural or unnatural at any given time within any given society. This book demonstrates that even so-called hard sciences, such as immunology, can profit from insights drawn from anthropology and sociology, and that the future of human milk studies needs to become genuinely transdisciplinary (Hassiotou et al. 2015). Whether milk is regarded as a biological resource, a nutritional necessity, or a symbol of relational exchange, a holistic research response is required. To understand human milk banking is to understand ties that bind, envisioning the maternal not only as a biological state but also as a strong yet flexible societal value and much of what it means to be human in a world that is ancient and modern at one and the same time.

The mission of anthropology is to make the strange familiar, and the mission of sociology is to make the familiar strange.[2] In either case, the constructed quality of that which is assumed to be "natural" emerges. Milk banks are fascinating from a theoretical point of view because they are at one and the same time "natural" and "scientific". Milk banks, especially research milk banks – expand the frontiers of so-called hard science while at the same time affirming (in the words of one milk bank manager) that "it's not rocket science". Milk itself is at one and the same time a known and unknown substance. There is no such thing as a "milk group", and for millennia, infants have relied on milk from someone other than their birth mothers in order to survive. Yet at the same time, the specific genetic properties of human milk, as encoded at a cellular level, are yielding ever newer discoveries. Historically, this paradox has been exploited above all by formula milk companies themselves, who have, strangely but logically enough, expanded the knowledge base of what is supposedly good, healthy and "natural" about human milk in the very effort to synthesize and supersede it.

At the end of this second decade of the twenty-first century, we are witnessing an exponential expansion of human milk health services occurring around the world, albeit without frequent or regular global discussions and agreements regarding international standardization and good practice (PATH 2013; DeMarchis et al. 2017; PATH 2017; Brandstetter et al. 2018). The term "bank" seems to be used internationally, although, as we will discuss again in greater detail, in some parts of the world, the commercial implications of this term are off-putting, and the term milk sharing is preferred (Daud et al. 2016; AL-Naqeeb 2000)—a usage which, however, risks confusion with very different practices of infant feeding not all of which are supported by healthcare communities. As we will also discuss, it is not coincidental that this growth has occurred at the same time that more informal online exchanges, either soliciting human milk as a gift donation or through what has been called "sharing", have also been expanding globally (Cassidy 2012a; Falls 2017; Palmquist and Doehler 2016). The ambition of this book is to link these human milk exchanges and the politics of breastfeeding to contemporary anthropological discussions of the politics of life itself (Rabinow 1996, 2007; Rose 2001, 2006; Rabinow and Rose 2006) and to notions of biopower, as it is evolving in connection with the expansion of human milk research around the world, a topic we discuss in Chapters 2 and 3, and the increasing entanglements between bioscience, biotechnological and economics, which Franklin and Lock (2003) term biocapital (see also Helmreich 2008), but which Waldby (2000) calls biovalue, which is "generated wherever the generative and transformative productivity of living entities can be instrumentalized along lines which make them useful for human projects" (Waldby 2000, 33; see also Waldby and Mitchell 2006). Underlying these discussions of the politics of life itself and the globally expanding bioeconomy are institutional links regarding the making and remaking of value for biological materials, especially when framed in terms of gender and fluidity (Krøløkke 2018), often part of the underworlds, and around which border struggles of power and inequity are expressed (Pavone and Goven 2017).

How we do or do not value human milk, and the mothers and their babies necessary to produce this substance, we argue, needs to have a broader anthropological vision of value itself. Furthermore, we argue for a study of the value of human milk within a larger economy of maternal corporeal generosity (Diprose 2002), which, as Shaw argues (2017) goes beyond altruism, usually defined.

Clinical considerations and medicalizing milk

From a clinical perspective, there has been increasing evidence accumulated for the clinical applications of an exclusively human milk diet, in particular for infants born prematurely, and especially in the context of evidence that bovine-based products have been shown to increase mortality and/or lead to poorer outcomes in terms of morbidity in particular from necrotizing enterocolitis (NEC) (Kantorowska et al. 2016; Feinberg et al. 2017). Although mother's own milk (MOM) is widely identified as the optimal first choice, and the choice we would certainly support, there has also been an increased recognition of the prevalence of delayed lactation, as well as other lactation difficulties, which a significant percentage of mothers of prematurely born infant(s) may experience (Meier et al. 2007; Meier et al. 2013). In short, the infants most urgently in need of mother's (or mothers') milk, are likely to be born to mothers least able to provide it. As we have discussed elsewhere (Cassidy and Dykes forthcoming) within the worlds of donor human milk, the milk is viewed relationally, as we argue a "liquid bridge" and not, as others have argued (Meier, Patel, and Esquerra-Zwiers 2016), as a simple replacement for formula. For the staff involved in the worlds of donor human milk services, as well as the mothers who donate, the milk is not a product, but a gift of self and is viewed within the relational support system for mothers who ideally will be able to increase their own production resulting in the first choice becoming possible for their own infant(s), and as we discuss in Chapter 6, ideally leads, in some cases, to these recipient mothers themselves becoming donors.

In 2010, the National Institute for Health and Care Excellence (NICE) prepared guidelines about donor human milk services across the UK which they recently offered in the form of an interactive flowchart form on their website.[3] NICE is an independent organization which provides guidance or national advice on the prevention and treatment of ill health and the promotion of good health (NICE 2005). The guidelines prepared for donor human milk were originally called "donor milk banks: the operation of donor milk bank services" but are now entitled "donor milk banks: service operations", demonstrating the range of responsibilities and sequence of operations involved in running a milk bank according to NICE guidelines. These functions involve the screening and selection of donors, training and support for donors and, finally, the processing of donor milk at the milk bank itself, including best practice safety guidelines. All of the milk banks involved in our study follow these guidelines and in turn these guidelines have been used to help establish similar guidelines in other countries around the world. These guidelines, however, are not so prescriptive as to preclude considerable procedural variation of practice among individual banks. Such banks typically have very few full-time staff and day-to-day practice is determined by the specific skills, experiences and priorities of a relatively small number of individuals. Furthermore, these guidelines do not cover issues associated with recipients, other than to discuss tracking procedures. During our data collection, the British Association of Perinatal Medicine (BAPM) published (2016) a framework for practice which said the following issues were not covered in the NICE guidelines, specifically related to the use of donor human milk (DHM) (BAPM 2016, 3):

a   The indications for the use of DHM, including definitions of at-risk groups that might benefit
b   What the benefits of feeding DHM are

c   The care and treatment of babies who receive DHM
d   How mothers should handle and store breastmilk for their own babies

The BAPM report discusses a telephone survey by Zipitis, Ward and Bajaj (2015) of all of the neonatal units across the UK, which found that use of DHM varied tremendously across the UK, with the Scottish system standing out with a universal coverage. However, large areas of (in particular) Northeast England are recorded as not using these services with cost cited as the main justification. Concluding that "there is currently inadequate evidence to make firm recommendations", the report went on to say that there is "evidence of efficacy and cost-effectiveness is urgently needed to determine the optimal indications for use and provision of DHM" (BAPM 2016, 2). The BAPM report also points out how extremely important it is for this service to be replacing the use of artificial feeds, not MOM's, which research continually indicates should always be the first choice. Whenever MOM is unavailable (usually for a short period of time), donor human milk has increasingly been recognized as the best practice second alternative. Moreover, increasing research from around the world, including most recently India (Adhisivam et al. 2017), indicates that donor human milk services can directly contribute to increasing exclusive breastfeeding rates in particular for the most vulnerable infants, a point we will return to in Chapters 4 and 5. The BAPM group met twice, and members included healthcare providers as well as managers from two of the milk banks involved in our research, both of which are presented as models for networks and hospitals across the UK to continue to develop their own policies and practices for use. However, with relevance to our research, the BAPM report, unlike the NICE guidelines group, did not have parental representation, nor did they present critical evidence for the studies they included.

Building liquid bridges

Milk banking also provides evidence that can help resolve a tension between "lactivism" (a portmanteau of lactation and activism) and feminism, arguing that these relations enable movement beyond "choice" (Smith, Haussman, and Labbok 2012). A gendered-contested term that originates with embodied experiences of mothering from the 1950s and the La Leche League (LLL), and is intimately tied to the professionalization of lactation consultants, but at the same time continues to be at the heart of debates regarding moralization of infant feeding (Dykes 2005, 2006, 2013). The term "lactivist" according to the *Oxford English Dictionary* is first used in 1999 and describes anyone who seeks to normalize breastfeeding in everyday environments, in particular public breastfeeding, thus de-sexualizing the female breast and re-emphasizing its nurturing function. While lactivists (many of whom might self-identify as feminists), seek to normalize breastfeeding in public spaces, some feminists recoil from lactivist rhetoric, which they feel seeks to overdetermine womanhood in terms of a very traditional function. The pressures that are placed on mothers to breastfeed their children may (and sometimes do) conflict with women's professional identity and sense of self. At a time when motherhood still carries significant professional penalties (Fuller and Hirsch

2019), lactivism threatens to force women back into the home and away from the workplace. But this is also a social movement run by women for women and may be seen as threatening to some, but as solidarity to others.

Evolved far beyond the "Breast is Best" mantra, twenty-first-century debates discuss logistical lactation challenges that many mothers experience, considering those challenges that are particularly complicated for mothers of preterm infants, some of whom are unable (at least initially) to provide sufficient milk to their infants, infants for whom human milk can be considered more medicine than nutrition. The "breastfeeding problem" was identified by healthcare providers over a century ago (Snyder 1908), dividing international expert opinion at the time between those who advocated "humanized" milk formula solutions (Rotch 1890) and those who sought the use of human milk but only under medically controlled procedures (Budin 1900, 1907) that eventually became known as donor human milk banking. This century-old health service has been expanding (albeit not consistently and not without setbacks) globally, often without critical and informed discussion, and limited socio-cultural research has been conducted on this provision.

To this extent, lactivism illustrates a theoretical division between (sometimes) competing feminisms that dates back many decades, a division that is generally felt to be represented by French and American schools of feminist thought. Is feminism about decentring masculinism and celebrating philosophy and praxis that celebrates distinctively feminized values or is feminism about escaping all traditional (imposed) definitions of femininity and seeking representation at every level of power on equal terms? This debate usually takes place within an environment which privileges a version of bourgeois individualism. It assumes, quite fairly for much of the time, that women have very little community support and that the feeding of infants is a uniquely individual choice and/or dilemma. An alternative model (familiar to much of the world today and the entirety of the world at various points in history) imagines a communitarian rather than an individual commitment to the feeding of infants – the so-called it-takes-a-village[4] model of parenting, the original title we considered for our final chapter, which in the end we call "liquid bridges". Such an alternative lactivism would try to imagine how workplaces and work regimes could be re-organized in order to prioritize human milk–fed babies as an absolute societal value. If human milk is acknowledged as being highly important, then the responsibility for facilitating it belongs to men as well as to women, and the ability to imagine different economic and ergonomic models becomes a moral and political imperative for everyone.

Ethnographic explorations

This book is based on the largest and most detailed comparative ethnographic research conducted by a senior researcher on donor human milk services and offers the most detailed systematic ethnographic information gathered on donor human milk services across the UK to date, which has been discussed and presented using international and interdisciplinary visions. We (Tanya and Fiona) first met each other in October 2009 when Fiona gave a keynote address to the Association of Lactation Consultants of Ireland (ALCI) entitled "Global Strategy for Infant and Young Child Feeding: The Rhetoric-Reality Gap" in Maynooth, Ireland, Tanya's home for over 20 years. Eventually, this led to our EU Horizon 2020 Marie Skłodowska-Curie Award (MSCA)[5] funded research entitled MUIMME (*M*ilk Banking and the *U*ncertain *I*nteraction between *M*aternal *M*ilk and *E*thanol), a two-year (2015–2017) multi-sited ethnographic study of the four largest donor human milk service clinics across the UK. We used a triangulated data collection approach, including observational fieldwork in each of the four banks each month for a year. Also, we gathered narrative interviews from staff, donors and parents of recipients from each bank. Thirdly, we collected documents, online and archival materials, for each of the four banks. In addition, Tanya also visited other services across the UK and in several parts of Europe and North America, having worked in this area for over 13 years.

Tanya has worked as a social scientist on gender and family health and nutrition issues for over 20 years, returning from a maternity break to work on the issue of human milk donation. Fiona has worked to develop an international voice for researchers, midwives and mothers (2006), and took on the role of sponsor, mentor, colleague and friend throughout this research. Tanya collected the data in this study, and when the first person is used in this book, it is Tanya who is speaking on her own; when we speak together we will use the term "we" to refer to both of us, although we worked closely together throughout this project and continue to collaborate on future projects. We both have backgrounds in ethnographic research, but from different perspectives: one as a social scientist and one as a maternal and child health services expert in ethnographic research, although we both recognize the voice of women, and in particular mothers, involved in the cultural world of MAINN. In addition, in Chapter 3, we have enlisted contributions from Professor Bernard Patrick Mahon, a world-renowned immunologist and Tanya's colleague in Ireland; therefore, in that chapter, he forms part of our joint voice. This is not, therefore, an ethnography in a traditional or classical anthropological sense, although we are discussing the cultural world(s) of donor human milk services, which is interlinked with cultures of breastfeeding, as well as with hospital cultures and reproductive health services. Our ethnography of the "contemporary" world of donor human milk exchange we envisioned as "a moving ratio of modernity, moving through the recent past and near future in a (non-linear) space" (Rabinow 2007, 2). This is one of many possible stories regarding the world(s) of donor human milk banking, and this is a story which is deeply indebted to the generosity of those who helped us with our MSCA project.

Moreover, our MSCA project and this book are situated in the new Feminist Anthropology frame, which is concerned not to ghettoize any given topic or area of research but instead, as Rayna Rapp (2016) has recently argued, pledges to make those things hidden in plain sight visible. People in the world of donor human milk services often hear people saying donor human milk bank service, "I never heard of that before", and sometimes people say, "If I had known when I was feeding my infant, I would have been thrilled to donate", and some even say that this would have kept them from having to pour what some call "liquid gold" down the drain. Having a recognition of social justice in one's study is key, and being inclusive, ethical and giving voice to the hidden are also defining, as we now go on to discuss.

Ethical concerns and considerations

An EU directive (Iphofen 2015) argues that ethnography, particularly from an anthropological perspective, should be considered as a form of continuing ethical decision making and that training in this form of ethical consideration should be a central part of all ethnographic research. Our study underwent extensive ethical reviews, both within the EU and as part of our funding receipt, and at the University of Central Lancashire, where we took on an ethics advisor (Dr Gill Thomson, reader/associate professor and member of the MAINN Unit) before making our extensive submission for ethical approval to the UK National Health Service (NHS) Health Research Authority (HRA) National Research Ethics Service,[6] including a face-to-face interview at which we were commended for our extensive and thorough submission, but were asked if we might not consider using the term "culture" instead of the term "ethnography". We discussed with the board the complexity of the term "culture" from an anthropological perspective, but acknowledging a lay interpretation, we included it on our information sheets, albeit in parentheses to acknowledge its contingent and complex understanding.

An important issue for many ethnographic narratives is related to concerns with anonymity and confidentiality, not only for those who we interview but also for the donor human milk banks sites themselves. Richard Titmuss' *The Gift Relationship* has long been acknowledged as one of the classic texts on social policy of human tissue exchange. Originally published in 1972, it takes blood donation as its primary focus, but Titmiss' daughter, Professor Ann Oakley, republished a new edition of her father's classic in 1997 with an additional chapter on donor human milk (Weaver and Williams 1997). Contrasting the British voluntary donors' system with the American one in which the blood supply can involve monetary compensation, he argues how a nonmarket system based on altruism (which was to evolve to include anonymity) is more effective than one that treats human blood, and by extension, milk, as another commodity (Strong 2009). The anonymity of the gift is an issue we will revisit in a later chapter. For now, it is important to remember, as Mills and colleagues (2010, 2) note, that anonymity and confidentiality are different in the following ways:

> Anonymity is the protection of a research participant's or site's identity. Confidentiality is the safeguarding of information obtained in confidence during the course of the research study.

As we will discuss, many people who participated in our study also publicly presented their stories, often with the hope that this would help donor human banking services to expand and in turn were sometimes reprinted on some of the websites associated with these services. Although we do mention the names of the sites involved in our study, and occasionally the names of some staff, and in one case the name of a donor, please note that all identifiable information is available in the public domain and therefore was not part of any confidential research collection.

Our ethnography was essentially four case studies, four donor human milk banks with the managers being active parts of the original study design and key collaborators throughout the data collection, although we are responsible for the narrative we offer the reader in this book. Each of the managers of these services was designated as our local collaborator and was treated as co-investigator for most of the project, offering co-authorship on selected articles with regard to their respective services. There is close collaboration and cooperation among donor human milk services across the UK, facilitated by the UKAMB. This was important since neither the Scottish nor the Northern Ireland site can be anonymized in any way, and therefore the anonymity of the sites involved in this study are not possible. But we are very clear that throughout our discussion, confidentiality is maintained and therefore only information which has been agreed to be made available is discussed. Moreover, regarding our narrative interviews, anonymity is maintained whenever possible, although some of these stories were subsequently retold in the public domain, which has to do with the saleable quality of many of these stories in terms of helping to expand knowledge about donor human milk service itself.

The four milk banks we studied were chosen because they were, at the time of our study, the largest across the UK and represented variability (two hospital-based and two community-based), as well as offering geographical coverage. As Figure 1.1 illustrates, we studied the only service on the island of Ireland, the *Milk Bank*, as it is called, a cross-border health cooperations from its inception,

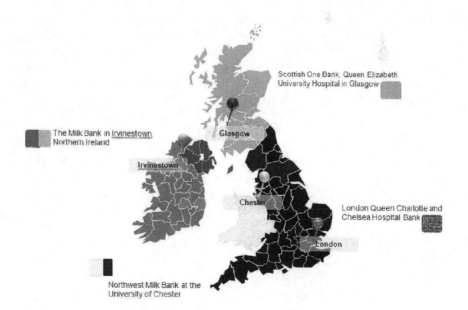

Figure 1.1 Map of MUIMME research sites in UK and Ireland

whose future may require redefinition following the UK decision to leave the EU (Brexit) following a national referendum that occurred during our data collection. This event is something we think is important to consider regarding the future of milk banking services internationally. Our second milk bank is the Scottish wide service, located at the Queen Elizabeth University Hospital in Glasgow. The third milk banking service was formed in 2013 following an amalgamation of the Countess of Chester and Arrow Park Hospital services, and the relocation to the NoW[7] Food Centre at the University of Chester is the *Northwest Human Milk Bank*, on the border of England and Wales, again offering a cross-border service from its inception (since Wales does not have its own milk bank, despite having the second-oldest service in the UK, as we will discuss in Chapter 2). The fourth milk bank included in our study is the QCCH bank in London, the oldest milk bank in the UK, located in the oldest maternity service, which we will detail more in Chapter 2.

Trust through space and time

We map the role of the hospital wet nurse, once widespread throughout the UK medical system, but highly criticized when associated with high rates of infant mortality, especially within so-called orphan or "foundling" hospitals as they were once variously known. The gendering of professional roles in healthcare settings affects the conflictual expansion of infant feeding for "weaklings" (as prematurely born infants were once widely known as in the medical world) whose notoriety was expanding globally. International differences associated with risk assessment and pasteurization[8] change following wartime concerns resulted in the establishment of Human Milk Bureaus in the UK, following a North American vision which eventually was to evolve into our modern system of banking on bodies, inextricably linked to blood banks, although organizationally predating and originating these medically controlled exchanges (Swanson 2014). Eventually, bureaus in the UK NHS system became known as "banks", which we describe in Chapter 2 using archival materials regarding the origins of the current four largest UK human milk banks.

The relations in the UK around human milk exchange were often devalued, as they were always gendered but empowering for the women involved in these exchanges. People like Edith Dare offer modernity for the care of the most vulnerable infants in wartime Europe. Despite rations, as well as other wartime constraints, and perhaps because of her philanthropic funding, she actively engaged technologies which would otherwise only be available to the wealthiest, including air transport and motorcycles for the delivery of milk, two technologies which continue to be utilized in long-distance transportation of this precious commodity, albeit on a voluntary basis.

Technologies of the science of milk

The origin narratives about donor human milk services are inextricably linked to clinical science, a discussion we then present in terms of non-linear links between contemporary breastmilk science and medicalized human milk services, not only in terms of the communal expansion of breastfeeding policies but also in relationship to the global expansion of human milk services in particular for the healthcare provision of prematurely born infants. For over a century, research has explored the medicinal and preventative qualities of breastmilk with reference to intestinal infections, such as NEC. Also included in this discussion will be the anti-viral and anti-bacterial properties of human milk, along with more recent scientific research endeavours to identify cancer-fighting properties in milk. In addition, researchers are exploring breastmilk as a potential source of stem cells (see for instance Witkowska-Zimny and Kaminska-El-Hassan 2017). It should be noted that chemical analyses have not always been in the service of a pro-breastfeeding agenda and have in fact been sponsored by attempts to synthesize various milk formulas. To posit breastmilk as "naturally" suited to the demands of vulnerable infants has often meant subjecting breastmilk to the most scientific and clinical of interventions. In turn, these scientific constructions have recently been applauded by lay communities of breastfeeding groups, arguing that the scientific frame removes moral undertones, a point we critically interrogate.

It was Engels rather than Marx who fully theorized the idea of the family in terms of exploitative class relations, opening up the possibility that families can be reconfigured when economic and political change permits (though perhaps not before). The family, according to this tradition, is less a biological essentialist norm than a contingent economic configuration. Some medical practitioners, however, have warned against the over-development of milk banks since donated milk cannot be regarded as advantageous as MOM, and the over-availability of donor milk might therefore distract mothers from a primary responsibility to learn how to feed their own babies. This rather harsh and punitive logic has been refuted by our own ongoing research, which suggests that the recipients of donor milk inhabit a "pro-breast" paradigm, are likely to want to do everything in their power to lactate and produce their own breastmilk and are inclined to sponsor breastfeeding-friendly environments more generally. This logic also reflects a rather narrow and individualistic view of the maternal function.

It is striking, meanwhile, that the concept of division of labour continues to decisively inform the marketing of donor milk, helping to promote the figure of the "good father" who is enabled, with the use of formula, to take on his "fair share" of responsibility for an infant allowing the mother to rest. However, as I was told during one of my early visits to one of the milk services involved in this study, and based on substantial personal experience, a regime of pumping creates plenty of work for fathers in the way of preparation, cleaning, storage and attending to an infant's need while the mother is immobilized by a pump.

Such theorizations would be obviously incomplete without an acknowledgement of the economics of breastfeeding as well as breast-donation (what might be termed "breastwork"). Despite what some well-intentioned lactivists may claim, breastfeeding is never "free" since nothing that involves the investment of time can be disentangled from large ergonomic and economic structures. Milk expression may also involve a problematic relationship with technology. As Marx declared in Volume 1 of *Capital* – machinery is never employed to "save labour" – but rather to increase surplus value – a proposition that is as true today as when he formulated it in the 1860s. Some women feel tyrannized by a regime of pumping and the pump itself becomes imagined as a stern employer. Feeling one with the pump may create a sense of cyborgification, which feels anything but liberating and evokes the image of Charlie Chaplin caught up in the cogs of the machinery he works with and for in *Modern Times*.

Translating transactions beyond gifts

In the UK, as well as many other parts of the world, the donation of expressed milk is not rewarded directly in monetary terms and is described as a free and generous "gift". Anthropologists do not regard a "gift" as some straightforward expression of pure and simple altruism, however, but as a symbolic action which makes sense in terms of larger communitarian norms and reciprocities. Most famously, Mauss (1925) considered the gift as something that creates bonds of trust and co-dependency that form the building blocks of social consensus. The gift becomes central to his larger concept of the "habitus", which has the effect of making practices that are culturally specific look as though they enjoy a biological inevitability. Those who give milk often speak of a sense of gratitude for the opportunity they have been given as well as identifying with a larger invisible community. There is also a temporal component to this sense of gifting community since there is a sense of wanting to "pay it forward" expressed as well as a strong sense of empathy with infants and parents of infants in need.

The translation of these issues is a key feature of the everyday life of a donor human milk bank. Organized around the definition of a milk bank, in Chapter 4, we discuss collection, screening, processing, storing and distribution of donor human milk every day. The working regime, the practical environment and the staffing of milk banks create very specific social environments and interactions which normalize particular pressures and priorities. By considering "time" and "space", two highly theorized concepts in the social sciences and the humanities, we offer interpretations of their shortage and shortfall within the worlds of human milk banking.

For many reasons, 2016 will forever be remembered. Not only was 2016 the year Donald Trump was elected president of the USA but also a very slim majority in the UK (but not Scotland, Northern Ireland or London, the site of three of our research sites) voted to leave the EU. For us, 2016 also marked the beginning of our data collection on which this monograph is based, which as we mentioned, followed over nine months of obtaining NHS HRA ethical approval, a process and experience which was to change significantly almost immediately after our experience and one we have also discussed elsewhere (Cassidy and Dykes forthcoming). Significantly for our discussion 2016 also saw a new translation of Marcel Mauss' classic essay *Essai sur le don: forme et raison de l'échange dans les sociétés archaiques* (1925), which Jane Guyer (Mauss 2016) has embedded into its original published frame, which recognizes for the first time its post-war and post-Durkheimian death associated with this original publication. Mauss, the nephew of Emile Durkheim, straddles the boundaries of both sociology and anthropology, and has, accordingly, been linked to the development of social anthropology in

Europe. This recent translation is argued to recover the ancestral links of this societal formations (Mauss 1925, 1954, 2002, 2016). It is itself a gift to the ancestors and therefore becomes part of those ancestors.

It is important to note that gifting does not represent a flight from the economic realm but rather a reconfiguration of economics in terms of a recognition of the versatility of networks of exchange. Such plurality has been theorized ever since the pioneering anthropological work of Bronislaw Malinowski (1922), whose theorization of the "kula ring" among Trobriand Islanders enabled him to explore an economic model which did not seek competitive advantage or surplus value but instead aspired to a kind of purity of reciprocal exchange. The persistence of such traditions into the twentieth century serves as a reminder that "the dismal science" of economics refers in the main to a Western industrial variant of the human experience and cannot be regarded as constitutive of what it means to be human, or a woman or a mother.

To understand milk donation, it is critical to understand both the pressure and the pleasure of milk donation. It is also a process which interrogates (sensitively and without cynicism) the essence of human generosity. Furthermore, milk donation suggests a redistribution of a primary nurturing function that Western industrial societies have too often imposed as a special and sacred ethical ambition on individual women.

Such utopian imaginings may appear far removed from practical policy-making in a broadly capitalist environment, but they serve to illustrate the fact that certain hegemonic neo-liberal assumptions are resisted by individuals and groups in surprising ways. In the meantime, however, there can be no doubt that "breastmilk" is commodified, insofar as it cannot help but be measured in cost terms. There is an internal market within the UK NHS which treats human milk as a commodity like any other, to be budgeted for and exchanged within a system that seeks to minimize losses. Within the far more complex and varied milk banking landscape of the US, donors are, on occasion, financially recompensed for their labour. Indeed, it becomes harder to insist rigidly on the supposed purity of the altruistic milk gift when everybody except the actual producer stands to gain economically from the product. Gifts have been traditionally theorized in oppositional terms to commodification, as was the case with Titmuss, but we espouse a more complex and synergistic understanding of "the gift" and commodity relations—one that considers the relations underlying the gift, including gendered relations.

Human milk is part of the human body. The body is, after all, composed mainly of fluids—many of which may be expelled on a regular basis. Psychoanalytic thinkers have long reflected on the implications for subject formation of various substances which can be expelled from the self and which the self has a problematic proprietorial claim upon, leading to various theories of "abjection" (Kristeva 1982; Kristeva and Goldhammer 1985). Kristeva has used the idea of womanhood and maternity to interrogate what is and is not sayable within a linguistic order that has figured as masculinist "symbolic" and oppositional. Her early work derived from her doctoral studies imagined a "revolution in poetic language", which illustrated how a semiotic "chora" (or field) of free-flowing potentialities precedes the phallocentric differential matrix of organized language and has never fully been repressed by it. Poetic language (temporarily) liberates a form of linguistic play that phallogocentric discourse seeks to contain. Her subsequent work, dominated by the famous *Stabat Mater*[9] (1985), theorizes the mythologizing of motherhood (with particular and very personal reference to Catholic Mariolatry) in terms of that which is and is not the self, and what one does and does not own. The ubiquitous separations that define this condition of maternity means that all maternity is grief as an abyss opens up between the offspring and the mother who was once indistinguishable from the child. In a talk Kristeva (2013) originally gave at MU, she discusses her personal experiences of being a mother, in particular the mother of a child with ongoing medical needs as a process which establishes matrescence as a permanently evolving condition.

HAU is an important open access ethnographic theoretical voice in social anthropology, which takes its name, as they say on their website,[10] from Marcel Mauss' important discussion of gift exchange in which he argues that something intangible of the giver remains part of the gift and is a Maori term meaning "spirit of the gift". In 2016, HAU, which also supported the retranslation of Mauss' work we mentioned earlier (2016), supported the publication of an important discussion regarding gender and theory by Marilyn Strathern (2016) with an introduction by Sarah Franklin and an afterword by Judith Butler under the title *Before and After Gender: Sexual Mythologies of Everyday Life*. All three of these women have intellectually contributed to our understanding of ethnography and of gendering the gift. Strathern's (1988) original discussion captures important post-modern feminist anthropological visions of exchange (Rosaldo and Lamphere 1974; Rapp 1975, including the article by Gayle Rubin) where gift giving is not seen as commodity exchange but rather as social relations, the complexity of which are key features of interaction. These visions help us to acknowledge and reconfigure understandings of maternal/corporeal (Diprose 2002) generosity in terms of a re-theorization of exchange, where the bodily exchange relationships can be viewed from the lens of becoming a mother, what Dana Raphael (1973) called matrescence,[11] a critical rite of passage, which involves both biological, social and cultural transformations, much akin to those experiences at other points in the lifecycle. Recently, Alexandra Sacks (2017; 2018), a reproductive psychiatrist, has presented this term in popular media, calling for a recognition of the identity changes inherent in maternity. As we discuss in Chapter 3, recent microbiological research in microchimerism argues that foetal maternal transference occurs not only for infants but also for mothers, and these cellular transferences are intergenerational (Kinder et al. 2017).

We explore how donor human milk banking involves establishing a community of generosity in Chapter 5 where we discuss exchanges between mothers, donors and parents of recipients. This chapter argues that mothers or recipient babies are not the passive beneficiaries of milk banking but have contracted into a network of relationships governed by a common sense of an urgent need for human milk for human babies. The motives, experiences and discoveries of donor and recipient mothers form the heart of this chapter.

Transitions and global expansions

As this exchange expands exponentially across the world, issues of equity of access, involving not only considerations regarding who receives milk but also in terms of who supplies the milk, are being discussed. We discuss these issues in relation to equity of access across the UK (which increasingly means access across borders and devolved healthcare systems), but also in terms of border crossings and the complexity of giving voice to mothers in this exchange. Comparative discussion will involve European and global examples making use of the special case of Brazil, which we discuss in greater detail in Chapter 6. As part of this equity of access, we need to think about where milk comes from and where it goes to, the nature of the exchanges between all women who are able to contribute and all children who are in need and whether this situation should remain in the control of a medicalized authority. International models of access will be part of our discussion. This is about the future of milk banking services across the world and how it is transforming at accelerating speeds. The central themes of trust, translation and technology are all components of this more global perspective and of our narrative throughout these maternal transactions. In addition, at the beginning of our research, the long-standing manager (over 25 years) of one of the services retired and then co-founded another service with another previous staff member, and although we did not specifically include this service in our study, this new service generated a great deal of publicity during our research and was often part of discussions during some of my ethnographic visits, and so we have included a discussion about transformations at the end of our discussion which includes a vision inspired by this bank with a difference and the potential future(s) of donor human milk services globally.

Throughout our discussion, we not only refer to the relational or interactional, as presented through both the social and the cultural aspects of donor human milk services, but also the maternal and the complex interplay involving the (micro)biological and those links to the medical frames underlying these exchanges. These transdisciplinary visions are expanding, along with the sciences of human milk, and potentially will eventually lead to new disciplinary productions, of which ethnographic visions will be central (Clifford and Marcus 1986). Further theoretical treatment of alternative models of economic exchange are consequent upon the development of online networking. Milk banking and milk sharing share an uneasy proximity since milk banks have always been keen (for the best of reasons) to assert the medical safeguards that come from a regulated storage and testing environment. However, the promotion of milk banking shares the same social media as milk sharing and understandable confusion therefore arises. Following Manuel Castells' (2000) pioneering work on the dawning "network society", milk sharing becomes a fascinating example of the ways in which the speed and ubiquity of communication creates alternative means of satisfying the most basic human need there is. These networks need not be hierarchical and may not be committed to generating surplus value (although they may be—in more general terms— highly profitable). A post-industrial world in which actual production forms a very small percentage of the dynamic economy fixates in creating new ways in which interested parties can socialize and communicate. The creation of milk communities represents, therefore, a twenty-first-century revivification of the pre-industrial "village" that was once regarded as central to healthy child rearing, albeit such "villages" consist of virtual and elective communities rather than physical and geographical spaces. Finally, human milk donation demands a theorized understanding of the sociological concept of "risk" as most influentially developed by Ulrich Beck (1992). Beck examines the structuring influence of competing risks— risks which cannot be eliminated and which need to be prioritized. Certain risks are preferred to others, not necessarily because of the absolute or quantifiable plausibility of the risk but because some risks can be assimilated more easily into a sense of what is regarded as consensually familiar and therefore acceptable. In a hospital setting, the concept of risk can, paradoxically, be especially risky, since the ethico-legal framework that defines much of hospital governance is more concerned with so-called sins of commission rather than sins of omission.

In short, the slightest possibility of a vulnerable infant being exposed to a contaminant via breastfeeding incurs greater anxiety than the much greater possibility of the infant failing to develop an immune system while being raised on formula (Boyd et al. 2006; McGuire and Anthony 2003; O'Connor et al. 2003; Schanler et al. 2005; Quigley 2007). Donor milk, because of its unfamiliarity, is considered a dangerous "intervention", whereas the normalization of bovine-based formula does not isolate it as a deliberate "risk" in the same way. It becomes impossible to separate the prevalence and extremity of risks from cultural contexts which privilege and stratify risks on the basis of whether or not they appear to form part of any "normal" course of events. When risks are normalized, they become invisible.

The range of theoretical traditions that can inform conversations about milk banking may seem dizzying. As dynamic and evolving traditions, they are themselves capable of adapting to absorb the complex evidential implications of milk banking as a worldwide practice. Milk banking exists, when it does exist, by becoming normalized, and the establishment of norms involves a far broader and more interdisciplinary imagination than a narrowly biomedical data set can hope to provide. In short, theoretical engagement is needed to naturalize the practice of milk banking because theoretical engagement is what determines the continued re-invention of the concept of "nature" itself.

Rubik cube of banking on milk

Our ultimate goal is to proffer a discussion of the everyday relations involved in donor human milk exchange. This detailed analysis helps to paint a picture of the complexity of interactions and could be represented by a network of relations, but instead we propose an alternative heuristic, a childhood mathematical toy, specifically the Rubik's cube.[12] Originally presented with six colours (red, blue, yellow, orange, green and white, meaning three primary and two secondary colours, and essentially a non-colour), argued by its inventor to be widely accessible to people around the world and offering potentially billions of potential combinations, and although it is ultimately finite, we suggest that it is the possibility of complexity of solutions which is most interesting.

Figure 1.2 Rubik's Cube of Banking on Milk

Transdisciplinary, trust, technology, translation, transition and transaction visually make up our combinatorial cubic narrative of the complex culture worlds of donor human milk banking (Ellis 1999; Ellis and Bochner 1999, 2000; Foster et al. 2006). Unlike the original instructions for the cube, we are not suggesting a search for a uniform solution, but instead argue for the complexity of possibilities offered by the journey, remembering the links between children and their mothers necessary to produce the milk at the centre of these exchanges while also acknowledging these exchanges include human milk service staff, other healthcare staff, as well as the extended families of both donors and recipients, creating and extending kinship relations through space and across time, all of which contributes to the relational cube of banking on human milk.

Notes

1 Rhonda Shaw (2004) first discussed a notion of the "yuk factor", a sense of something that tastes bad, and you might spit out, whereas "ick" captures the visceral nature of disgust which can be associated with milk from another mother, something one would never consider putting in one's mouth.
2 This is a widely used paraphrase of what has been argued to date back to the late eighteenth century German poet, Georg Philipp Friedrich Freiherr von Hardenberg who wrote under the pseudonym Novalis (1772–1801) translated in modern times as "to romanticize the world is to make us aware of the magic, mystery and wonder of the world; it is to educate the senses to see the ordinary as extraordinary, the familiar as strange, the mundane as sacred, the finite as infinite" (Beiser 1998: 294). This is interesting because many of the early Chicago sociologists and anthropologists would have had strong German influences.
3 There is controversy about when and how this popular phrase came into use. There are links with West Africa and the publication by Jane Cowen-Fletcher (1994) of the same name, although it was popularized by Hilary Rodham Clinton (1996). We use it here to capture the collaborative collective nature of the exchange involved in donor human milk banking.
4 Project ID 654495.
5 REC ID 15/NW/0762; IRAS #181994.
6 Please see https://pathways.nice.org.uk/pathways/donor-breast-milk-banks and note that we have received permission to re-produce this screen image, which was taken on 31 May 2018.
7 NoW stands for northwest, as in the region of the UK.
8 Bruno Latour (1988) has talked about this in relationship to France but did not include any discussion of human milk in his book.
9 At the bottom of this essay, it states this essay was first published "under the title 'Herethique de l'amour', in Tel Quel, no. 74 (Winter 1977). It was reprinted under the present title in Kristeva, Histoires d'amour (Paris: Denoel, 1983). This is the first English translation, very slightly shortened from the original. Translated by Arthur Goldhammer" (1985: 133).
10 See www.haujournal.org/index.php/hau/.
11 Raphael says this term was "capped" by Professor Conrad Arensberg (Raphael 1973: 19). Arensberg's ethnographic work on Ireland (Arensberg 1937, Arensberg and Kimball 1940, 1965, 2013) not only formed his own doctoral work but also was groundbreaking for ethnographic study of industrial nations and continues to shape the anthropological studies of Ireland and Europe in general. Both he and Kimball helped to found the Society for Applied Anthropology (SfAA) contributing to discussions of culture, policy and practice throughout their careers.
12 Rubik's Cube® used by permission Rubik's Brand Ltd. Please note a colour version of this image can be obtained by emailing Tanya.Cassidy@mu.ie.

References

Arensberg, Conrad. 1937. *The Irish Countryman*. New York and London: Macmillan.
Arensberg, Conrad and Solon T. Kimball. 1940. *Family and Community in Ireland*. Cambridge, MA: Harvard University Press.
———. 1965. *Culture and Community*. New York: Harcourt, Brace and World.
———. 2001. *Family and Community in Ireland*. 3rd ed. With a New Introduction by Anne Byrne, Ricca Edmondson, and Tony Varley. Ennis. County Clare: CLASP Press
Adhisivam, B., B. Vishnu Bhat, N. Banupriya, Rachel Poorna, Nishad Plakkal, and C. Palanivel. 2017. "Impact of Human Milk Banking on Neonatal Mortality, Necrotizing Enterocolitis, and Exclusive Breastfeeding – Experience from a Tertiary Care Teaching Hospital, South India." *The Journal of Maternal-Fetal & Neonatal Medicine* 32(6): 902–5.
BAPM (British Association of Perinatal Medicine). 2016. "The Use of Donor Human Expressed Breast Milk in Newborn Infants A Framework for Practice." www.bapm.org/sites/default/files/files/DEBM%20framework%20July%202016.pdf
Beck, Ulrich (translated by Mark Ritter). 1992. *Risk Society: Towards a new modernity*. London: Sage Publications.
Beiser, Frederick C. 1998. "A Romantic Education: The Concept of Bildung in Early German Romanticism". In *Philosophers on Education: Historical Perspectives*, ed. Amélie Oksenberg Rorty. London and New York: Routledge: pp. 284–99.
Brandstetter, S., K. Mansen, A. DeMarchis, N. Nguyen Quyhn, C. Engmann, and K. Israel-Ballard. 2018. "A Decision Tree for Donor Human Milk: An Example Tool to Protect, Promote, and Support Breastfeeding." *Frontiers in Pediatrics* 6: 324.
Budin, Pierre. 1900. *Le nourrisson: alimentation et hygiène – enfants débiles, enfants nés à terme*. Paris: O. Doin.
———. 1907. *The Nursling. The Feeding and Hygiene of Premature and Full-Term Infants*. London: Caxton.
Cassidy, Tanya M. 2012a. 'Mothers, Milk, and Money: Maternal corporeal generosity, sociological social psychological trust, and value in human milk exchange.' Special Issue on Motherhood and Economics. *Journal of the Motherhood Initiative (JMI)* 3. 1: 96–111.
Cassidy, Tanya and Fiona Dykes. Forthcoming. "Building Liquid Bridges with Donor Human Breastmilk."
Castells, Manuel. 2000. *The Rise of The Network Society: The Information Age: Economy, Society and Culture, Volume 1*. Cambridge, MA: Blackwell Publishing.
Clifford, J. and G. Marcus. 1986. *Writing Culture: The Poetics and Politics of Ethnography*. Berkeley, University of California Press.
Cowen-Fletcher, Jane. 1994. *It Takes a Village*. New York: Scholastic Press.
Daud, Normadiah, Nadhirah Nordin, Zurita Mohd Yusoff, and Rahimah Embong. 2016. Chapter 46 "The Development of Milk Bank According to Islamic Law for Preserving the Progeny of Baby." In *Contemporary Issues and Development in the Global Halal Industry*, edited by Siti Khadijah Ab. Manan, Fadilah Abd Rahman, Mardhiyyah Sahri. Syngapore: Springer.
DeMarchis, A., Kiersten Israel-Ballard, Kimberly Amundson Mansen, and C. Engmann. 2017. "Establishing an Integrated Human Milk Banking Approach to Strengthen Newborn Care." *Journal of Perinatology* 37: 469–74.
Diprose, Rosalyn. 2002. *Corporeal Generosity: On Giving with Nietzsche, Merleau-Ponty, and Levinas*. Albany, NY: SUNY Press.
Dykes, F. 2005. "'Supply' and 'Demand': Breastfeeding as Labour." *Social Science & Medicine* 60: 2283–93.
———. 2006. *Breastfeeding in Hospital: Mothers, Midwives and the Production Line*. London: Routledge.

———. 2013. "Review: Militant Lactivism? Attachment Parenting and Intensive Motherhood in the UK and France by Charlotte Faircloth." *Sociology of Health & Illness* 35: 1128–129.

Ellis, Carolyn. 1999. "Heartful Autoethnography." *Qualitative Health Research* 9 (5): 669–83.

Ellis, Caroline and Arthur Bochner. 1999. "Bringing Emotion and Personal Narrative into Medical Social Science." *Health* 3: 229–37.

———. 2000. "Autoethnography, Personal Narrative, Reflexivity: Researcher as Subject." In *Handbook of Qualitative Research*, edited by N. K. Denzin and Y. S. Lincoln. 2nd ed., 733–68. Thousand Oaks, CA: Sage.

Falls, Susan. 2017. *White Gold: Stories of Breast Milk Sharing*. Lincoln: University of Nebraska Press.

Feinberg, M., L. Miller, B. Engers, K. Bigelow, A. Lewis, S. Brinker, . . ., J. R. Britton. 2017. "Reduced Necrotizing Enterocolitis After an Initiative to Promote Breastfeeding and Early Human Milk Administration." *Pediatric Quality & Safety* 2 (2). http://journals.lww.com/pqs/Fulltext/2017/03000/Reduced_Necrotizing_Enterocolitis_after_an.2.aspx

Foster, Kim, Margaret McAllister, and Louise O'Brien. 2006. "Extending the Boundaries: Autoethnography as an Emergent Method in Mental Health Nursing Research." *International Journal of Mental Health Nursing* 15 (1), March: 44–53.

Franklin, Sarah, and Margaret Lock, eds. 2003. *Remaking Life and Death: Toward an Anthropology of the Biosciences*. 1st ed. Santa Feq, Oxford: School for Advanced Research Press.

Fuller, Sylvia, and Hirsh, Elizabeth. 2019. "'Family-Friendly' Jobs and Motherhood Pay Penalties: The Impact of Flexible Work Arrangements Across the Educational Spectrum." *Work and Occupations*. 46(1): 3–44.

Helmreich, Stefan. 2008. "Species of Biocapital." *Science as Culture* 17: 463–78. doi:10.1080/09505430802519256

Iphofen, Ron. 2015. *Research ethics in ethnography/anthropology*. European Commission, DGResearch and Innovation.

Kantorowska, Agata, Julia C. Wei, Ronald S. Cohen, Ruth A. Lawrence, Jeffrey B. Gould, and Henry C. Lee. 2016. "Impact of Donor Milk Availability on Breast Milk Use and Necrotizing Enterocolitis Rates." *Pediatrics*, 2015–3123.

Kinder, J. M., I. A. Stelzer, P. C. Arck, and S. S. Way. 2017. "Immunological Implications of Pregnancy-Induced Microchimerism." *Nature Reviews. Immunology* 17 (8): 483–94.

Kristeva, Julia (translated by Leon S. Roudiez). 1982. *The Powers of Horror: An essay on abjection*. New York: Columbia University Press.

———. 2013. "A Tragedy and a dream: Disability revisited." *Irish Theological Quarterly*. 78(3): 219–30.

Kristeva, Julia, and Arthur Goldhammer. 1985. "Sabat Mater." *Poetics Today*, The Female Body in Western Culture: Semiotic Perspectives. 6(1/2): 133–52.

Kroløkke, Charlotte. 2018. *Global Fluids: The Cultural Politics of Reproductive Waste*. New York: Berghahn Books.

Latour, Bruno. 1988. *The Pasteurization of France*. Cambridge, MA: Harvard University Press.

Malinowski, Bronislaw (1922, second edition 1932). *Argonauts of the Western Pacific*. London: George Routledge & Sons, Ltd.

McGuire, W., and M. Y. Anthony. 2003. "Donor Human Milk versus Formula for Preventing Necrotising Enterocolitis in Preterm Infants: Systematic Review." *Archives of Disease in Childhood. Fetal and Neonatal Edition* 88: F11–F14.

Meier, Paula, A. L. Patel, K. Wright, and J. L. Engstrom. 2013. "Management of Breastfeeding During and After the Maternity Hospitalization for Late Preterm Infants." *Clinics in Perinatology* 40 (4): 689–705. doi:10.1016/j.clp.2013.07.014

Meier, P., A. Patel, and A. Esquerra-Zwiers. 2016. "Donor Human Milk Update: Evidence, Mechanisms, and Priorities for Research and Practice." *The Journal of Pediatrics* 180: 15–21.

Meier, P., L. M. Furman, and M. Degenhardt. 2007. "Increased Lactation Risk for Late Preterm Infants and Mothers: Evidence and Management Strategies to Protect Breastfeeding." *The Journal of Midwifery & Women's Health* 52 (6): 579–87.

Mills, A. J., Durepos, G., and Wiebe, E. (2010). Encyclopedia of case study research. Thousand Oaks, CA: SAGE Publications, Inc.

National Institute for Health and Clinical Excellence. 2005.

NICE (National Institute for Health and Clinical Excellence). 2005. *Social Value Judgements: Principles for the Development of NICE Guidance*. www.nice.org.uk/media/873/2F/SocialValueJudgementsDec05.pdf

O'Connor, D. L., J. Jacobs, R. Hall, D. Adamkin, N. Auestad, M. Castillo, W. E. Connor, S. L. Connor, K. Fitzgerald, S. Groh-Wargo, E. E. Hartmann, J. Janowsky, A. Lucas, D. Margeson, P. Mena, M. Neuringer, G. Ross, L. Singer, T. Stephenson, J. Szabo, and V. Zemon. 2003. "Growth and Development of Premature Infants Fed Predominantly Human Milk, Predominantly Premature Infant Formula, or a Combination of Human Milk and Premature Formula." *Journal of Pediatric Gastroenterology and Nutrition* 37 (4), October: 437–46.

Palmquist, A., and Doehler, K. 2016. Human Milk Sharing Practices in the US. *Maternal and Child Nutrition*, 12(2), 278–90.

PATH. 2017. "Policy Brief – Ensuring Equitable Access to Human Milk for All Infants: A Comprehensive Approach to Essential Newborn Care." www.path.org/publications/files/MNCHN_EquitableAccesstoHumanMilk_PolicyBrief.pdf

———. 2013. *Strengthening Human Milk Banking: A Global Implementation Framework*. Version 1.1. Seattle, Washington, DC, USA: Bill & Melinda Gates Foundation Grand Challenges initiative, PATH.

Pavone, Vincenzo and Joanna Goven. 2017. *Bioeconomies: Life, Technology, and Capital in the 21st Century*. Cham, Switzerland: Palgrave Macmillan.

Quigley, M., G. Henderson, M. Anthony, and W. McGuire. 2007. "Formula Milk versus Donor Breast Milk for Feeding Preterm or Low Birth Weight Infants." *Cochrane Database of Systematic Reviews*, October 17.

Rabinow, Paul. 1996. *Essays on the Anthropology of Reason*, Princeton: Princeton University Press.

———. 2007. *Marking Time: On the Anthropology of the Contemporary*. Princeton: Princeton University Press.

Rabinow, P. and N. Rose. 2006. "Biopower Today." *BioSocieties* 1: 195–217.

Raphael, Dana. 1973. *The Tender Gift: Breastfeeding*. Englewood Cliffs, NJ: Prentice-Hall, Inc.

Rapp Reiter, Rayna (ed.). 1975. *Toward an Anthropology of Women*. New York: Monthly Review Press.

Rapp, Rayna. 2016. "Prologue." In *Mapping Feminist Anthropology in the Twenty-First Century*, edited by Ellen Lewin and Leni M. Silverstein. New Brunswick, NJ and London: Rutgers University Press.

Rosaldo, Zimbalist Michelle and Louise Lamphere, eds. 1974. *Woman, Culture, and Society*. Stanford, CA: Stanford University Press.

Rose, N. 2001. "The Politics of Life Itself." *Theory, Culture & Society* 18 (6): 1–30.

———. 2006. *The Politics of Life Itself: Biomedicine, Power and Subjectivity in the Twenty-First Century*. Princeton, NJ: Princeton University Press.

Rotch, Thomas Morgan. 1890. "The Management of Human Breast-Milk in Cases of Difficult Infantile Digestion." *American Pediatric Society* 2: 88–101.

Sacks, Alexsandra. 2017. "The Birth of a Mother." *New York Times*. 8 May.

———. 2018. "A new way to think about the transition to motherhood." TED.com

Schanler RJ, Lau C, Hurst NM, et al. 2005. Randomized trial of donor human milk versus preterm formula as substitutes for mother's own milk in the feeding of extremely premature infants. *Pediatrics*. 116:400–6.

Shaw, Rhonda (ed.). 2017. *Bioethics Beyond Altruism Donating and Transforming Human Biological Materials*. Cham, Switzerland: Palgrave Macmillan.

———. 2004. "The Virtues of Cross-Nursing and the Yuk Factor." *Australian Feminist Studies: Special Issue on Cultures of Breastfeeding* 19 (45): 287–99.

Smith, P. H., B. Haussman, and M. Labbok, eds. 2012. *Beyond Health, Beyond Choice. Breastfeeding Constraints and Realities*. New Brunswick: Rutgers University Press.

Snyder, J. Ross. 1908. "The Breast Milk Problem." LI (15): 1212–14.

Strathern, Marilyn. 1988. *The Gender of the Gift Problems with Women and Problems with Society in Melanesia*. Berkeley: University of California Press.

———. 2016. *Before and After Gender*. Edited with an Introduction by Sarah Franklin Afterword by Judith Butler. Chicago: HAU Books.

Strong, Thomas. 2009. "Vital Publics of Pure Blood." *Body & Society* 15 (2): 169–91.

Swanson, Kara W. 2014. *Banking on the Body: The Market in Blood, Milk, and Sperm in Modern America*. Cambridge, MA: Harvard University Press.

Waldby, Catherine. 2000. *The Visible Human Project: Informatic Bodies and Posthuman Medicine*. London: Routledge.

Waldby, Catherine and Robert Mitchell. 2006. *Tissue Economies: Blood, Organs, and Cell Lines in Late Capitalism*. Durham, NC: Duke University Press.

Weaver, G. and A. S. Williams. 1997. "A Mother's Gift: The Milk of Human Kindness." In *The Gift Relationship*, edited by Titmuss, R. M., A. Oakley, and J. Ashton. 2nd ed., 319–32. New York: The New Press.

Witkowska-Zimny, M. and E. Kaminska-El-Hassan. 2017. "Cells of Human Breast Milk." *Cellular & Molecular Biology Letters* 22, 11. http://doi.org/10.1186/s11658-017-0042-4

Zipitis, C. S., J. Ward, and R. Bajaj. 2015. "Use of Donor Breast Milk in Neonatal Units in the UK." *Archives of Disease in Childhood – Fetal and Neonatal Edition* 100: F279–F281.

2 Moving hospital wet nurses to bureaus and banks

Anthropologists have argued that allomaternal nursing, or nursing an infant from another mother, forms part of most cultures across the world and throughout history (Hewlett and Winn 2014). It continues, however, to be an under discussed topic, and discussions of donor human milk services are also rare. As we discuss throughout this book, this is a reflection on the lack of value placed on women and the feeding of infants, an issue which seems to be changing globally (WHO 2003). Historians often concentrate instead on wet nursing (Sussman 1982; Apple 1987; Fildes 1988; 1986), although that sometimes has included a discussion of the medicalization of these services (Golden 1996; Appadurai 1986; Marx 1977). Also, seeking to redress the perceived lack of value placed on human milk research, Kara Swanson (2014) offers a more recent historical discussion on the "banking" of bodily fluids. Although human milk is included in her discussion, blood is discussed in greater detail, a bodily fluid which one could argue has a much more widely recognized social value as "therapeutic merchandize", a phrase used originally in 1929 by an American physician to describe the use of human milk from a new form of "foster mother" (Tobey 1929, 1110), a phrase Janet Golden (1996) adopted to describe the entire organization of human milk in the twentieth century:

> In 1900 wet nurses occupied several small niches—suckling foundlings in institutions or working for well-to-do private families. By the 1910s and 1920s the number of wet nurses in these venues had decreased, although new opportunities arose for women willing to suckle abandoned babies in their homes or premature infants in hospitals. At the same time, a new career opened for lactating mothers: expressing and selling their breast milk for use in homes and hospitals. This procedure proved so successful that by the 1930s wet nurses had almost entirely vanished, replaced by bottled human milk. As one physician described it, human milk had become "therapeutic merchandise".
>
> (Golden 1996, 179)

In the UK, there have been fewer historical studies, with the notable exception of A. Susan Williams study (1997) of the UK National Birthday Trust, and a separate chapter Williams co-authored with Weaver (1997) which appeared in a reissue of Richard Titmuss' *The Gift Relationship* (1997), which was co-edited by his daughter Ann Oakley, a widely recognized sociological professor of maternal studies. Co-incidentally, Titmuss' book was recently reissued (October 2018), and although Oakley points out in an editorial preface that much of the 1997 version was used in this latest edition, with the 1997 additional materials including the chapter on "the gift of human milk" were not included (2018: v). Also, co-incidentally, 1997 marks the establishment of the UKAMB.[1]

Sue Balmer, the long-time manager of the Sorento Maternity Hospital milk banking service in Birmingham which opened originally in 1950 at the Children's Hospital and moved to Sorento Hospital in 1955 (Balmer and Wharton 1992; Wharton 1981), wrote a historical discussion of UKAMB, which was privately published through UKAMB (Balmer 2010). Although Balmer had previously published some historical discussion (Balmer and Wharton 1992), this more extensive UKAMB publication had a limited distribution,[2] a copy of which Tanya was given, while another copy was also available for reference in the office of one of the milk bank sites. We also visited the archives at the Wellcome Trust Library as well as the British Library and any archives for each of the four milk bank services. Using materials gathered throughout our ethnographic work, we wish to present a brief historical overview of the four donor human milk services involved in study, which we have organized around some of the key historical topics. Beginning in England, which is the home of the oldest donor human milk service in the UK, we will discuss how human milk services are linked to the much older practice of "wet nursing", but in particular the hospital wet nurse. These early organized wet nursing provisions, particularly in London, involved women who had experienced crisis pregnancies and whose infants often ended up in foundling hospitals, many of which were Irish, so it is logical that we will move on to discuss the human milk services in Ireland, which seems to have had a historically older tradition and more culturally integrated system associated with human milk services, although as we will discuss, this system was almost completely erased by the influx of artificial feeding. We will then discuss the situation in Scotland, which also almost completely lost its culture of breastfeeding and was one of the key first services to be re-introduced in the 1970s (Slimesand Hallman 1979), and which survived the onset of HIV/AIDS in the 1980s (Anonymous 1988a, 1988b) to eventually become a Scottish wide service. Finally, we will consider services for Wales, which seems to have been the first service in the UK to use the term bank, although today there are no services available in Wales or in the Republic of Ireland, while both regions, as we will discuss, have been served by one of the largest banks involved in our study, and in some ways one of the youngest: the service in Chester.

The canning of medicalized mothers' milk

It is widely recognized that donor human milk services, although they were not called banks yet, began in 1909 in Vienna (Jones 2003), and accordingly in 2009, there was celebratory conference titled "100 Years of Milk Banking" organized by members of the EMBA in Vienna, Austria (Cassidy 2009).[3] As critical researchers, we felt compelled to ask the questions why 1909, and why Vienna? As a result, I discovered (Cassidy 2009) that there was important international research being done among the newly forming profession of paediatricians in Vienna, which was linked to the team under the already renowned Theodor Escherich, whose work on intestinal bacteria led to him giving his name to e-coli (Mayerhofer and Přibram 1909a, 1909b, 1912).

On 5 September 1909, the *Washington Post* (p. M1) reports under the heading "Canned Mother's Milk" that "Dr. Mayerhofer, of Vienna is making experiments with preserved (canned) woman's milk. Results in Francis Joseph's Child Hospital show that preserved mother's milk serves much better than the cow's as first infant food. A new coming strange industry" (1909: M1).

That same month (1909), two articles had been published by Ernest Mayerhofer and Ernest Přbram, both in German, one under the title "Feeding Experiments with Conserved Women's Milk (Frauenmilch)" and the other "On Food including Canned (or Conserved) Woman's Milk (Frauenmilch)" and the *British Medical Journal* (1909, 1005) also published the following:

> IN a communication made to the Vienna Medical Society Drs. Mazerhofer [sic] and Przibram [sic], of Professor Escherich's paediatric clinic, reported the results of feeding newborn babies with preserved human milk. The milk is drawn with a specially designed breast-valve-pump. It is possible to obtain all the milk contained in the gland by its action without the least discomfort to the woman. Then the milk is, for purposes of storage, sterilized by means of the Budde process (heating to 55°C. in the water bath, with an addition of a few grams of hydrogen peroxide). The milk thus treated is completely sterile, without any alteration in its nutritive value, taste, or appearance, as reported by Escherich and many other observers. The authors have fed forty infants with this milk; the babies throve just as well as the babies of the mothers from whose breasts the milk was obtained. The breasts seem to adapt themselves to the increased demand to which they are subjected. The advantage of this process lies in the fact that the milk can be drawn off whenever a wet-nurse is present and can be kept until needed. As there are daily about 120 suckling women in the Vienna lying-in hospitals, and each application of the pump easily yields three ounces, a quantity of about 120 pints could be had daily for the feeding of babies of mothers who cannot suckle them. The matter was taken up by some daily papers here, and precautions have been demanded to prevent the method being made a form of commercial enterprise. The only way of preventing mischief is to restrict the use of the method to hospitals for infants. The lives of many babies born before full term, hitherto carefully but only with difficulty reared in the *couveuse*, will in future be much more easily saved, as the keeping qualities of the milk are practically unlimited.

It is reasonable to conclude that Drs. Mayerhofer and Přibram's research marks the first time human (mother's) milk was pasteurized and then redistributed to infants in hospital care. The facilities allowed this milk to be preserved and used as a first infant food, which was understood to be better than the commercial bovine counterpart. Affiliated with the children's department of the Francis Joseph's Hospital (Kaiser-Franz-Josef Spital) in Vienna, as we said, one of the most important centres for developing paediatric specialism (Anonymous 1902, 1926; Kepler 1988;

Cassidy 2009). In 1902, Theodor Escherich took over the management of the University Hospital and St Anna Children's Hospital, and with public funds established the University Children's Hospital, which opened in 1911 shortly after his unexpected death (Kepler 1988). As part of this work, Escherich pioneered maternal education and counselling in his new hospital setting through his "Nursing Station" (Kepler 1988). In a third article, published a year after Escherich's death, Mayerhofer and Přibram (1912) said they had been working on this method since 1908 in order to reduce the problems associated with not having access to a wet nurse. This article also reports that similar results have been found at another "nursing station" (Thiemichschen Säuglingsstation) in Magdeburg, which is interesting in the history of milk banking, since this is where the first German "women's milk collection point" was established by Marie-Elise Kayser (1885–1950) in 1919 (Seifert 2012).

Vienna had an extremely strong and unionized profession of wet nurses, as is evidenced by the highly publicized strike of wet nurses in Vienna in 1894 (*British Medical Journal* 1894). We know that on both sides of the Atlantic, hospital wet nurses were routinely employed to feed hospitalized infants (Budin 1907; Silverman 1979), and from the later part of the nineteenth century into the early part of the twentieth century, these nurses were captured as part of the public display of premature babies on both sides of the Atlantic. The extremely high rate of mortality, particular among infants in foundling hospitals, which medical authorities attributed to them not receiving a human milk diet, which recent evidence supports (Arthi and Schneider 2017). The question of how to provide these infants human milk was linked to the availability of wet nurses, who were also paid and, in some countries, professionally organized, while in other countries, these women became increasingly linked to exploitation of either their own or other people's infants. The medical control of wet nurses, including taking control of how and where wet nurses can be chosen, became increasingly common and is linked not only to the human milk service which was established in Boston at the Floating Hospital in 1910 by Dr Francis Parkman Denny (1869–1948), who had himself completed post-graduate training in Vienna and Berlin at the end of the nineteenth century (Golden 1988, 1996), which undoubtedly fuelled his recognition of the importance of breastmilk for infant health. At the same time, in the UK, this type of service is also linked to Queen Charlotte's Lying-In Hospital and the eventual birth of human milk services across the UK.

SOS for human milk

Queen Charlotte's Lying-In Hospital in London, one of the oldest maternity units in the UK, had long advertised in popular newspapers, including the *Times* (1888) and the *Observer* (1889) "wet nurses" would be provided "promptly" "upon application to the matron". On 3 December 1938, the *Lancet* announced that the "National Mothers' Milk Bureau will be opened by the National Birthday Trust Fund at the new Queen Charlotte's Hospital, Hammersmith, on Jan. 1st, 1939" (Lancet 1938, 1307):

> It is announced that a National Mothers' Milk Bureau will be opened by the National Birthday Trust Fund at the new Queen Charlotte's Hospital,

Hammersmith, on Jan. 1st, 1939. The bureau is to be under the supervision of the hospital staff, but the cost of maintenance will be borne by the fund. Through the generosity of Sir Julien Cahn, chairman of the fund, the most modern equipment is being provided. The object is to make human milk available for delicate babies for whom it has been medically prescribed. Details of price and methods of preservation and of guaranteeing the health of the donors are not yet available, but it is probable they will be largely based on the methods now used at the ten human-milk centres in the United States of America.

(p. 1307)

Sir Julien Cahn, labelled an "eccentric and well-travelled British millionaire" (Weaver and Williams 1997, 327), was also the director of the National Birthday Trust Fund.

There seems to be a certain amount of confusion regarding exactly when the Queen Charlotte's Lying-In service actually began, as the hospital was moving from its previous long-term home in Marylebone to its modern-day home in Hammersmith. On the 2 January 1939, the *Daily Herald* reported that the day before, the "National Mothers' Milk Bureau" opened the "life-saving bureau for babies". On 6 January, *The Cambridge Independent Press* and *Chronical* (1939, 6) reports under the banner heading "Safer Motherhood":

> One of the most congenial engagements of the Queen before she goes to Canada will be her visit to Guildhall to call national attention to the "Safer Motherhood" campaign of the National Birthday Trust Fund.
>
> Sir Julien Cahn, generous friend of hospitals and of Notts County cricket, is chairman of the fund and grand helper.
>
> Noble work is being done but the need for development is great. The latest development is the opening this week of a national Mothers' Milk Bureau established in a brank of Queen Charlotte's Hospital. It is equipped on most modern lines in order to provide human milk for babies for whom the doctors consider it a vital necessity. The Birthday Trust has undertaken to maintain the bureau.

On 1 March 1939 in the *Daily Herald*, an article, along with a series of pictures said to have been taken by E. G. Malindine, gives a pictorial display of the entire process of the human milk bank. The article says, "This 'Bureau' will be opened on March 14" (Calder 1939, 20). Copies of these images were later given to Queen Charlotte's and placed in their historical materials, although their original origins were later lost. The article has a total of ten images, a row of three, a row of four, then a row with two images and below that a single image, beginning with the image of a woman expressing her milk using a water pump.[4] We were unable to purchase the original images from this article, but we found additional images taken on that same day that were available through Historic England who kindly gave us permission to reprint them here (Figures 2.1–2.4).

The caption under the image of filtering in the *Daily Herald* says, "It is filtered through eight layers of gauze into a chromium-plated cylinder", and the second

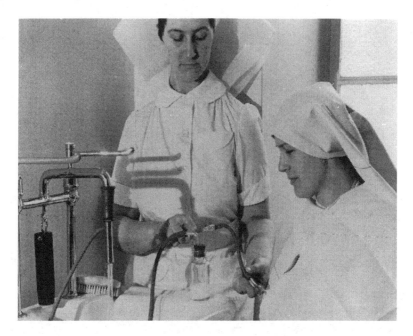

Figure 2.1 Using a water pump the milk is drawn out and put into sterilized glass containers.
Source: Reprinted with permission from Historic England.

Figure 2.2 We are told that the milk is filtered through eight layers of gauze into a chromium-plated cylinder.
Source: Reprinted with permission from Historic England.

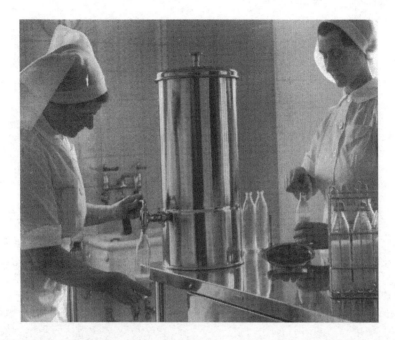

Figure 2.3 After filtering the milk we are told it is poured into specially prepared feeding bottles.
Source: Reprinted with permission from Historic England.

Figure 2.4 The bottles are placed in a basket which are then placed in the pasteurizing machine.
Source: Reprinted with permission from Historic England.

image had the caption "from which it is poured into specially prepared feeding bottles". Historic England also has images of the bottles being washed, which we have not included here, as this is something only alluded to in the original *Daily Herald* article where it says the bottles are specially prepared.

This image has the caption, "These are placed in a pasteurizing machine and are kept at a temperature which destroys every risk of germs or contamination". The next image, which we do not have a copy, shows the bottles being placed in a refrigerator with the caption "After being pasteurized the bottles are then stored in special refrigerator, where they can safely be kept for 48 hours" showing the same person placing the basket of bottles of pasteurized milk into a refrigerator (Figure 2.5).

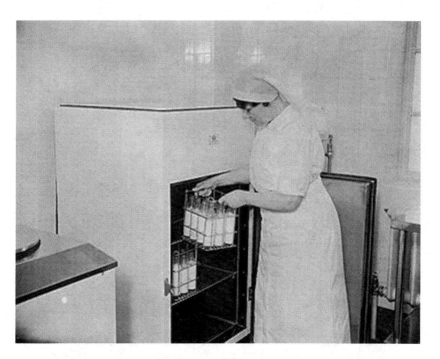

Figure 2.5 After being pasteurized the milk is stored in a refrigerator for up to 48 hours.
Source: Reprinted with permission from Historic England.

The next image (similar to the earlier one, but not the exact one) had the caption, "If the daily demand for the milk is less than the supply, it is then placed in these biscuit-shaped moulds before being frozen". The image we have shown earlier has the same glass tube which was used to fill the moulds (Figure 2.6). This next image (the forth in a row) has the caption, "The moulds are sandwiched between blocks of carbon dioxide, which freeze the milk solid at a temperature of 10 deg. below freezing point" (Figure 2.7).

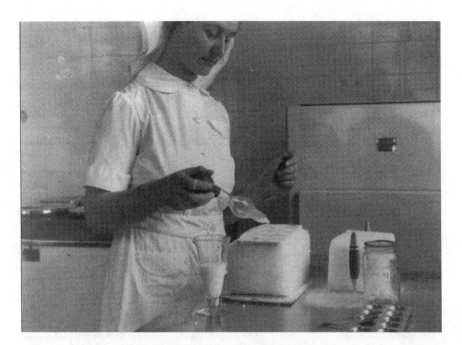

Figure 2.6 If the daily demand is less than the supply we are told that the excess milk is placed in biscuit-shaped moulds before being frozen.
Source: Reprinted with permission from Historic England.

Figure 2.7 The biscuit-shaped moulds are placed between blocks of carbon dioxide, which freeze the milk solid at a temperature of 10 degrees below freezing.
Source: Reprinted with permission from Historic England.

32 *Moving hospital wet nurses*

The article continues and shows the milk being frozen into small discs. These discs are then placed in a glass jar and which is stored in a cold box ready to be defrost when needed. The newspaper image does not show the human milk on the side, but instead crops the image so as to just show the ice (Figures 2.8 and 2.9).

Under both these two images together is the caption, "These frozen tablets, weighing one third of an ounce, can be kept, if necessary, in refrigerators for six months without losing any vitamin value which cannot be replaced by fresh fruit juices".

The last image, a copy of which was not available in Queen Charlotte's materials nor Historic England, shows four nurses riding bicycles and has the caption, "Daily supplies of this human milk are delivered where needed by nurses, who are seen leaving one of Queen Charlotte's Homes". The transportation of the milk is something that Edith Dare and the team at the Queen Charlotte's milk service were very proud of, and later we see images of nurses using motorized bicycles for a variety of things.

Later that same month, the *Daily Mirror* newspaper has an article talking about "one of its newest branches" of the National Birthday Trust and goes on to say it "is the provision of a human milk bureau, which has been made in co-operation with Queen Charlotte's Hospital" (Ascroft 1939, 11). This story continues, in bold print, to say, "This new service is expected to save the lives of many delicate babies whose own mothers are unable to feed them" (Ascroft 1939, 11). The reporter finishes by saying that she feels "the most important thing of all, is the peace of mind

Figure 2.8 We are told that the frozen tablets weigh one third of an ounce each.
Source: Reprinted with permission from Historic England.

Figure 2.9 The frozen tablets are placed in glass jars and then in a refrigerator.
Source: Reprinted with permission from Historic England.

brought by the knowledge that expert medical assistance is there if they need it, and that they will not have to bear unnecessary agony" (Ascroft 1939, 11).

Later that year, in June 1939, a much more detailed report is published in the *British Medical Journal*, which states, "The bureau was set up partly as a result of the necessity for supplying human milk to the St Neot's 'quads'[5] and in answer to an SOS messages received from time to time" (*British Medical Journal* 1939, 1298). Thanks to links made by healthcare providers, the first of which was this anonymous article in the *British Medical Journal*, the celebrated UK quadruplets became a part of the origin story of the human milk bureau despite being born four years before it was to officially open.

We reviewed the British Newspaper Archive and discovered that several newspapers reported that in October 1935, another set of quadruplets (all boys) were born in London. A number of British newspapers followed these children, two of whom, unfortunately do not survive, the first dying shortly after being christianed, and the second one dying in November 1935 just before the second set of quads (one girl and three boys) were to be born at No 13 in Eynesbury, St Neot's Huntingdonshire (*Sunderland Daily Echo and Shipping Gazette* 1935, 3).

The St Neots Quads story is central to media coverage of early human milk services in the UK. The infants', the second group of four in eight weeks, were delivered at home (a council house, as the father is a lorry/truck driver) and attended by Dr Ernest Henry Harrison on his own. In the *Lancashire Post* that day, it reports that

13 must be their luck number, as that was the family's house number and the doctor's phone number. In this article, we are told that not only were the children born at home but also that they were not expected for another two months, so they were born prematurely. We are also told that they were fed sterilized water. We are then informed that this is the second set of quadruplets within the last two months, and then the story turns to discuss the four boys born the previous month, two of whom died, adding that having two sets of quadruplets in England within two months of each other is "very unusual" and that quadruplets are 2,000 times as rare as twins. The article ends by reminding the reader that the "famous Dionne quintuplets" (Anonymous 1935, 5) are now "18 months old and can walk unaided".

The following day, 29 November, several more newspapers report the birth of these four infants, but also add that they were being cared for in an improvised cot of "brown hide easy chairs" and that they are now "almost famous", reporting that a news reel of the babies was also being planned (Anonymous 1935, 5). And the newspapers report that Dr Harrison made "the quadruplets and their mother his first call of the day", and he says,

> "I am delighted to say that all four babies are doing exceptionally well and are very much improved today. Mrs. Miles too, is very comfortable, and her conditions affords me the greatest satisfaction. I have not decided as to what diet the babies will have, and the meantime they will continue to have sterilized water".
> (Anonymous 1935, 5)

This article says that the Dr Harrison's wife received so many cards that she was "almost as proud as the quadruplets' mother". The article then reports the quadruplets mother as saying,

> I am feeling so well that I would like to get up and lend a hand as the babies must be taking an awful lot of looking after, but there is not much hope of my being allowed to do this. I would like my little babies to be up in this room beside me. I have seen them—they are lovely.
> (Anonymous 1935, 5)

In the article, the grandmother is also quoted as saying that the naming of babies will not be done until the mother is "stronger", but this seems to have happened fairly quickly since the article ends with a discussion of the babies being christened, saying, "The girl was named Ann, the first boy Ernest, the next Paul, and the youngest Michael" (5). The mother not having power in this situation seems very clearly presented in this article. Not only is the physician seen as being in control, but the grandmother is also exercising her control over affairs, although her wishes seem to be somewhat superseded.

On 2 December, we learn that "quads cost £15 per day" because they are receiving "human milk from London" (*Western Morning News*, 7). The article goes on to say, "Human milk is brought by car from a London hospital at 10am and 6pm, the journeys totalling about 200 miles a day". "Ann takes her food ravenously, Paul well, Ernest fairly well, and Michael falls asleep after a teaspoonful, so he is given a teaspoonful whenever he wakes" (*Western Morning News*, 7).

The article tells us that the babies were given cots and that four nurses had been employed to help with their care, but also that they had been moved to their physician's home. The article ends by noting that the father was making the journey to and from London. That same day, the *Nottingham Evening Post* in an article titled "Daily Milk Supply by Aeroplane?" we are told, "Daily air trips to supply the babies with sterilized human milk from London are being contemplated by Mrs. Winifred Crossley, a daughter of Dr Harrison" (1). The article tells us she is an experienced pilot, and although she recently sold her plane, she is thinking about hiring a plane to make the daily trips. A month later, on 21 December, the *Sphere* publishes contrasting images of the babies, one with their father surrounding the babies wrapped and placed in the arm chairs and the other with their physician and mask-covered nurses attending the babies who are now placed separately in their own cots. These are the first images in the *Sphere* linked to "human milk" which are not about artificial alternatives.

As we know, all four of the St Neot's quadruplets survived, and later, we are told that their physician wrote the matron Edith Dare at Queen Charlotte's hospital in 1938, encouraging her to expand the service she had provided to the quadruplets on a large-scale, saying, "They never would have survived had it not been for you supplying, day after day, the proper quantity" (Paterson 1938). However, the *British Medical Journal* article from 1939 also says that a year or two before (so 1937 or 1938) Dr Leonard Colebrook visited Boston and "saw a similar bureau in operation, and on his return suggested that a service of this nature might be appropriately, run from Queen Charlotte's" (BMJ 1939, 1298). The article goes on to say that a "similar service" had been available for 25 years in the US and that there were organizations of this kind in Canada, Germany and Russia. It then also says that Miss Dare also went to Boston and received training on how to run the service, which they then describe in the following detail:

> The milk is obtained from nursing mothers who are recommended to the bureau by the medical officers of health of five West London boroughs. It is taken only from nursing mothers who have carried their infants to full term, not from any cases of prematurity or stillbirth, and the mothers have; to reach a high health standard. Before being placed on the bureau's list the milk is thoroughly tested. The mother is then supplied with the necessary equipment, consisting of a simple water pump, towels for covering hair and clothes, bottles in which the milk is stored, and a container of dry ice in which the bottle remains until collected. Fresh bottles are supplied each day after washing and sterilization at the bureau. The collection of the milk by car takes place each morning, the mother and her child are seen every day, and if the child is not making normal progress and attending an infant welfare centre once a week the milk is not accepted. Complete. records of the weight and progress of the children, and of the amount.
>
> When the milk reaches the bureau, it is tested in the laboratory for fat content, for any possible adulteration by illegitimate additions of water or cow's milk, and for dirt and bacteria. The bottles are then emptied into a container, the milk being strained through sterilized butter muslin as a further precaution and mixed together by a plunger. The milk is then poured into

fresh bottles and placed in a pasteurization plant, where it is heated to 145° F. for half an hour. After rapid cooling it is ready for use. Milk which is not immediately needed is placed in a refrigerator at 45° F. and left for twelve hours. At the end of the day unused milk is frozen into small cakes, about the size of a half-crown, between blocks of chemical ice (carbon dioxide ice), which are stored in glass bottles at a temperature of 10° F below zero. These frozen cakes retain all their vital constituents; on testing after six months' storage the difference between the stored and fresh milk has been found so slight as to be negligible. The bureau has at present twelve mothers on its books, with a total production of 150 to 180 oz. per day, and since its inception it has dealt with about 9,000 oz. of milk. It has supplied Great Ormond Street Hospital and some of the L.C.C. hospitals, also private practitioners. The mothers are paid at the rate of two pence per ounce. The improvement in their health and also the increased amount of milk produced is said to be remarkable. In some cases the financial assistance saves them from the necessity of going out to work, or is a means of providing them with extra nourishment. The bureau (Riverside 1126) is at call night and day, and the milk can be dispatched almost immediately to any part of the country. A sum of sixpence per ounce is charged, but special arrangements can be made in the case of necessitous patients.

Later that year, in October 1939, we are informed in a *British Medical Journal* article that a decision to evacuate expectant mothers was attempted but that mothers would not leave. The article goes on to discuss several hospitals, beginning with Queen Charlotte's, the largest hospital, which left its new premises in Hammersmith and returned to their old Marylebone Road site, incidentally the site of today's St Mary's hospital, which continues to be linked to both Queen Charlotte's and to the milk bank. The discussion ends by saying,

"One unhappy result of the closing down at Hammersmith is that the human milk bureau has had to go out of action temporarily. This was a service of national importance, and the hospital is now hard put to it to supply human for desperately ill babies".

(*British Medical Journal* 1939, 772)

However, this service was to open again very quickly, and we know the bank answered an "SOS" for human milk in Slough, which was delivered from London by car on the 29 April 1940 (*Daily Mirror*, 30 April 1940). Unfortunately, the following evening, we learn that despite "the consignment" being sent "[w]ithin a few minutes" from receiving the "broadcast", the 2lb 6oz infant died (*Evening Despatch*, 30 April 1940).

In the early part of the twentieth century, it was far more common for mothers to give birth at home. Hospital births were often linked to crisis pregnancies and were used for teaching purposes. But as the century progressed, the hospital became increasingly linked to the more medicalized or serious births, such as multiples or complicated births, a stable of obstetrics for several centuries (Porter

1987). As hospital births increased in popularity, there were huge implications for breastfeeding rates and experiences, including these famous quadruple births in St Neots, although we discover that within a month, the infants were taken into care in the physician's home to allow the physician and the nursing staff to have better access in more affluent settings; similar events occurred in Canada, albeit under more controversial circumstances.

Donald Paterson, who we have just mentioned, noted in the seventh edition (1939) of his textbook co-authored with J. Forest Smith entitled *Modern Methods of Feeding in Infancy and Childhood* that "the practice of wet nursing had been in serious decline" (Paterson and Smith 1939, 27). First published in 1926, this widely read textbook included a discussion of wet nursing as an "ideal complementary feed" saying,

> The ideal complementary feed be that of human milk, obtained from some mother who is secreting more than is necessary for her own infant. This is given to the infant who is not gaining immediately after it has been to its mother's breast. Wet nursing, where the infant gets the whole of its supply from the foster-mother, is not so popular in this country as it deserves to be, largely owing to lack of suitable foster-mothers. With care in selection, and after a preliminary inquiry into the health of the foster-mother, and the obtaining of a negative Wassermann reaction, this practice may at times be the only method of successfully rearing a weakly infant.
>
> (Paterson and Smith 1939, 27)

As we can observe, the authors are already lamenting that this tradition is no longer popular in the UK, despite the need for such provision. Subsequent editions of this popular textbook appear in the 1940s and 1950s (including the eighth (1945), ninth (1947) and tenth (1955) editions), but there are no longer any references to "wet nursing" in the later editions (Weaver and Williams 1997). We would contend that it is not coincidental that this time period also corresponds with the beginning of donor human milk services in the UK, as well as with the increasing use of artificial foods.

As we will discuss, donor human milk services continue to expand across the UK, while at the same time, human milk is being de-privileged by the formula industry, privileging "humanized" fluids, as is evidenced by a major advertisement from Trufood Limited, which appeared on page two of the *Times* in 1954.

When we investigate this advert, we discover that Bartholomäus Metlinger was a German physician who wrote one of the first textbooks with illustrations about the care of children, which was originally published under the title *Kinderbüchlein* ("Little Book on Children") on 7 December 1473, being retitled in later editions as *Ein Regiment der jungen Kinder* (A Guide on Young Children), several of which have been republished through the centuries and are available online and at the Wellcome Trust Library. Throughout the centuries, as Cassidy (2015) discusses elsewhere, there were many medical discussions regarding how to choose the best wet nurse, if needed, although certainly from the end of the eighteenth century onwards, mothers in many European cultures were actively encouraged

How to choose a Wet-Nurse

"*Where the milk does not come in or for other reasons the mother cannot nurse her child, one should choose a wet-nurse who has the following appearance and habits. She must be neither too young nor too old. She should be well-built; her face healthy in appearance, tanned; and she should have a strong thick neck, strong, broad breasts, not too fat and not too thin. The wet-nurse should have good, praise-worthy habits. She should not be easily frightened or worried and not small-minded or prone to anger.*"

BARTHOLOMAEUS METTLINGER (ABOUT 1450)
IN 'THE REGIME OF YOUR CHILDREN'

Times have certainly changed. Nowadays the equivalent of the wet-nurse is probably the breast-milk bank. Or, as another perfectly good substitute for breast-feeding, the modern mother can turn to Trufood. Here is a range of balanced infant foods which includes Humanised Trufood, the milk preparation nearest to breast milk. Trufood helps modern babies to grow into healthy, robust children who wouldn't know a wet-nurse if they saw one.

The Creameries and Laboratories of

TRUFOOD

LIMITED

are at Wrenbury in Cheshire

Figure 2.10 Reprinted with permission from the Times Digital Archive. Trufood Limited. *The Times* (London, England), Saturday, August 21, 1954, Issue 53016, p. 2.

to feed their own infants, who were much more likely to survive if fed MOM. (It should also be noted that the advert quotes the wrong year.) Also, the note to this advertisement says, "Nowadays the equivalent of the wet-nurse is probably the breast-milk bank". It goes on to say, "The modern mother can turn to Trufood", "another perfectly good substitute for breast-feeding", it is "nearest to breast milk" and it "helps modern babies to grow into healthy, robust children who wouldn't know a wet-nurse if they saw one".

Milk Wafers Latest Food for Hungry Babies

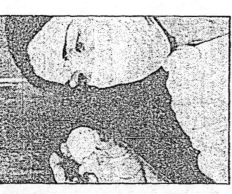

A nurse capping a bottle containing the raw milk.

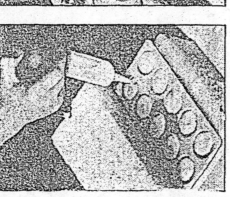

The pasteurised milk is poured into dry ice moulds.

Frozen in wafer form, the milk is ready for shipment.

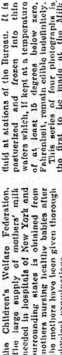

YEARS ago, when a new-born infant could not be fed by its mother, the doctor's first problem was to find a wet nurse. Since then, medical science has made gigantic strides, and the over problem of feeding hungry infants has been met in modern fashion by the establishment of the Mother's Milk Bureau of the Children's Welfare Federation. The large supply of mothers' milk needed in hospitals of New York and surrounding states is obtained from mothers nursing healthy babies after the mothers have been given thorough physical examinations.

The mothers who have this surplus milk supply contribute the precious fluid at stations of the Bureau. It is pasteurized and frozen into thin wafers which, if kept at a temperature of at least 15 degrees below zero, Fahrenheit, remain edible indefinitely.

This series of four photographs is the first to be made at the Milk Bureau and affords an interesting human insight into the latest advance in the care of babies.

Reprinted with permission from Independent Newspapers.

For the protection, promotion and support of breastfeeding

Queen Charlotte's human milk service in 1939 was not the first time the Irish and British public were to learn about a "Mothers' Milk Bureau", which was featured in the Irish *Sunday Independent* on 26 September 1937 (p. 4).

This story relates to the New York Mother's Milk Bureau of the Children's Welfare Federation in New York. Clearly, the paper emphasizes the need for this milk while pointing out that it is also being used in "surrounding states". To distance itself from any potential negative association with wet nursing and the exploitation of infants whose mothers supply this milk, it is pointed out that the milk "is obtained from mothers nursing healthy babies" who have undergone "physical examinations". This so-called precious fluid is then pasteurized and frozen into "milk wafers", employing key scientific processes which we will discuss more in the next chapter.[6]

The long history of wet nursing, particularly among Irish immigrant mothers in both London and America, which is ultimately derived from the ancient tradition of milk fostering in Ireland, is another reason why it is not surprising that the first images of a human milk bureau in this part of the world are published in the Irish *Sunday Independent* newspaper. As a matter of interest, the term "MUIMME" is not only the acronym we used for our EU MSCA–funded project (Milk Banking and the Uncertain Interaction between Maternal Milk and Ethanol) but is also an ancient Irish term for wet nurse[7] as well as foster mother, an association which is also common in many other cultures (Koch 2006; Cassidy 2015). However, as Cassidy (2015) discusses elsewhere, for over 200 years, the modern Irish term for wet nurse is "bean-chioch" (O'Reilly and O'Donovan 1817), although the most recent use is "banaltra chíche",[8] which literally translates as "woman nurse breast". Also, not unrelated to this, is the fact that the oldest maternity hospital in the world is the Rotunda hospital in Dublin, which was founded as a teaching hospital for so-called male midwives in 1745 by Bartholomew Mosse (Kirkpatrick and Jellett 1913), with research on human milk dating back to at least the 1780s (Clarke 1790), as was discussed by Joseph Clarke, master of the Rotunda between 1787–1793, who discussed his experiments, which he describes in the following fashion.

> Accordingly, I put this question to all our experienced nurse-tenders in the Lying-in Hospital "Is there any difference of colour in the curds vomited by infants of four or five days old and by those of a fortnight or three weeks?" It happened that two or three of them were sitting together when I first thought of proposing this question. They answered unanimously, and without hesitation, "Surely, Sir, there is; until the beesting milk is over the curds are yellow, and afterwards they become white".
>
> (Clarke 1790, 81)

Reviewing the annual reports for the Rotunda, we see that in 1934 in Appendix C that it says, "Breast milk has been our sheet-anchor for feeding the premature infants; where it has been impossible to obtain the mother's milk, a breastmilk pool has been made from other mothers willing to contribute" (Solomons 1934,

380), a report that predates the occasion of the St Neots quadruplets born in England and the famous deliveries of human milk sent from London. Like the largest lying-in hospital in London, the oldest lying-in hospital in Ireland was also an active user of human milk, in particular for infants born prematurely, but they also seem to be pooling the milk, although they are not pasteurizing it yet.

However, it seems, like in England, the term "bank" is not actively used until somewhat later; specifically, the earliest we were able to trace comes from a 1958 *Clinical Report for the Rotunda* (p. 19) which reports from Nurse Dolan.

> The Breast-Milk Bank has had a reasonably satisfactory year, collecting a total of 2,078 pints. The best month was December (264 pints), and the poorest, February (123 pints).
>
> In spite of all the propaganda and encouragement few mothers breast-feed for longer than six weeks, and a considerable number have never started or have given up before this time. Mothers who become donors will usually continue to lactate for 4–6 months, during which time they feed their own infants only, donating the excess to the Bank. Donors are hard to come by; few mothers being interested. The donor panel at any one time seldom exceeds 16, and not infrequently falls to as low as four donors.

In less than a decade, it is reported in a Galway paper that it was impossible to obtain milk from the Rotunda milk bank, reporting,

> A doctor said it was regrettable that breast feeding was now virtually extinct in Ireland, and was at its lowest incidence ever. In America, where the system of artificial feeding for infants had first been developed on a large scale, there was now a return to breast feeding, but there was no sign of a similar movement in this country.
>
> (Anonymous 1966, 1)

As had occurred with the more famous multiple births in Canada (Kiriline 1936; Berton 1977; National Library of Canada 2003; Rosack 2004), the St Neots infants were approached by the formula industry and ended up advertising artificial infant foods. Although there is certainly a place for the use of artificial feeding, and its use increased throughout the UK, it is instructive to be reminded of the fact that the Irish dairy industry is one of the world's largest producers of "infant nutrition products, producing 15% of the world's powdered infant formula" (Irish Business and Employers' Confederation 2007; More 2009). Throughout the 1950s and 1960s, as the earlier quote indicates, the expansion of the formula industry is pervasive, not only in Ireland but also across the UK (Schulman 2003).

After the stork breastmilk bank

In an interesting transcript of an expert witness seminar from the Wellcome Trust (which Dykes was a participant) concerning the resurgence of breastfeeding between 1975 to 2000 (Crowther, Reynolds, and Tansey 2009), we see historians arguing that the term resurgence may not be appropriate and that the rates had been dropping since the 1920s across the UK and Ireland. The introduction to the published account of this seminar written by Rima Apple refers to the intergenerational changes between two popular infant feeding texts, one from the nineteenth and early part of the twentieth century (Holt 1901) that argues that almost all mothers should nurse their own infants. She points out that Holt's son authors a 1957 popular childcare book which remarks in a non-committal fashion: "Breast or bottle? This is something that every mother must decide for herself" (Crowther, Reynolds, and Tansey 2009). The end of the 1970s and the beginning of the 1980s also saw the joint WHO UNICEF statement:

> Where it is not possible for the biological mother to breastfeed, the first alternative, if available, should be the use of Human Milk from other sources. Human Milk Banks should be available in appropriate situations.

In the UK, as Balmer (2010) points out, milk banking services continued to expand at this time and that the Department of Health and Social Services in 1981 listed 19 milk bank services across the UK, including now defunct services in Barnsley Bristol, Cardiff, Crewe, Dundee, Liverpool, Leinster, an additional service in London, and Manchester and Scunthrope. The list also includes several services still in existence today, including Queen Charlotte's services, as well as the service at the Queen Mother's Hospital (QMH) in Glasgow, which opened in 1978 (Figure 2.11).

Figure 2.11 Reprinted with permission from One Milk Bank for Scotland manager Debbie Barnett.

A recent discussion Scottish government online discussion of the donor milk bank said the service

> had no dedicated staff at this time and handled milk from 8 to 12 donors annually. This milk was pasteurised by the Yorkhill Milk Kitchen and the process was overseen by a Consultant Neonatologist and the Neonatal Unit Liaison Midwife.

Table 2.1 Donor Human Milk Banking in Scotland 2008–2018

Year	Litres Pasteurized	Donors	Recipients	Litres Distributed
2008	103	35	32	–
2009	317	42	84	–
2010	264	45	89	Moved
2011	428	64	104	–
2012	453	77	144	190
2013	635	100	194	435
2014	791	159	206	551
2015	914	153	324	700
2016	1,174	156	459	912
2017	1,199	204	525	1,056
2018 to April	435	94	209	418

Source: NHS Greater Glasgow and Clyde (2012). www.wired-gov.net/wg/news.nsf/articles/Funding+for+breast+milk+bank+23062018070500?open

Balmer (2010, 32) tells us that the consultant was neonatologist Dr Forrester Cockburn, and through the 1990s and the early part of the 2000s, the service was run by Rhonda Robinson. In a round-table discussion of international perspectives on human milk bank services, we are told that

> [i]nitially, the bank only provided milk to babies at the QMH and the co-located children's hospital (Royal Hospital for Sick Children). Recipients included preterm infants and those following various gastrointestinal surgical procedures, whose mothers could not produce sufficient milk. In 2008, the bank was expanded to provide milk to all maternity units within the Greater Glasgow and Clyde Health Board area, and due to increasing requests for donor milk from across Scotland, we are currently expanding further to provide a Scotland-wide service.
>
> (Grøvslien et al. 2013, 310)

The current manager took over in 2009, and in 2010, after the closure of the Queen Mother's hospital, the service moved to its current location. The service continued to expand, and in 2013, it became the Scottish wide service, as the picture of the plaque from the bank indicates.

Moreover, in June 2018, the Scottish government additionally pledged £300,000 towards the expansion of the bank, which is seen to be an integral part of the government's plans to support breastfeeding in general. Table 2.1 shows how the Scottish bank has expanded in the last five years.

Precious milk for precious babies in the twenty-first century

Following the discovery in mid-1980s of the potential contamination associated with the discovery that the HIV/AIDS could be transmitted from mothers to infants via their milk (Ziegler et al. 1985; Acheson 1988; Dunn et al. 1992), which dealt a death blow to many milk banks across the globe were going to receive in the 1980s (see Cassidy 2013). In 1992, there were only 11 milk banks still open across the UK, including Queen Charlotte's Hospital in London, as well as one at King's College Hospital and St George's Hospital in London, one in Birmingham (Sorrento Maternity Hospital), Cambridge (Rosie Maternity Hospital), Chatham (All Saints Hospital), Chertsey (St Peter's Hospital), Kingston on Thames (Kingston Hospital), Oxford (John Radcliffe Hospital), Southampton (Princess Anne Maternity Hospital) and one in Glasgow (Queen Mother's Hospital). This meant that there was now only one in Scotland and none in either Wales or Ireland, a situation which continued until the beginning of the twenty-first century, at least in Ireland. However, in response, regulations were being offered as working party set-up guidelines, which eventually formed the basis of the UK UKAMB in 1997 (Balmer 2010), with the founding services noted in Table 2.2.

Part of the remit of UKAMB was to hold roadshows and to discuss how services could be established and run, and in October 2000, one of these roadshows was held in Belfast and eventually led to the establishment of the service in Northern Ireland (Balmer 2010).

Across the island of Ireland all of human milk services had disappeared by the 1990s, with the only service in existence today opening in 2000, purposely near the border so as to service the island of Ireland as a whole and a direct health service provision made available thanks to the so-called Good Friday Agreement and the extension of cross-border cooperation (McCrea 2007; Cassidy 2015). Like the Scottish service, from its inception, the Irish human milk service has been linked to the expansion of breastfeeding rates generally in the community. Having the service run by someone trained in community health nursing who also actively supported community mothers with infant feeding problems has been key to this expansion, although it often meant that individual staff members have more than one job and therefore stretching the scope of the job, as we will discuss in Chapter 4. Historically, the closure of banks was linked to extremely low rates of breastfeeding in Ireland, and although these rates have been rising, they continue to some of the lowest in the world, and therefore there continue to be concerns

Table 2.2 Founder human milk banks

Birmingham	1 Birmingham Maternity Hospital
Cambridge	2 Rosie Maternity Hospital
Chatham	3 Medway Maritime Hospital
Chertsey	4 St Peter's Hospital
Glasgow	5 Queen Mother's Hospital
Huddersfield	6 Huddersfield Royal Infirmary
Kingston	7 Kingston Hospital
London	8 King's College Hospital,
	9 Queen Charlotte's Hospital
	10 St George's Hospital
Orpington	11 Princes Royal University Hospital
Oxford	12 John Radcliffe Hospital
Southampton	13 Princess Anne Maternity Hospital

Adapted from Balmer 2010, 46.

regarding a lack of available people able to support these services. Ireland was one of the earliest services to argue that donor human milk services can and should be linked to policies for improving breastfeeding rates across Ireland, an issue that is becoming recognized more and more internationally (DeMarchis et al. 2017; Xavier de Meneses et al. 2017; Adhisivam et al. 2017), as we will discuss more in Chapter 6, but also in relationship to donor human milk services in Scotland.

For many years, "The Milk Bank", as it is now simply called, was the only community-based human milk service across the UK, located in community health services in Irvinestown, Enniskillen, while linked to the South West Acute Hospital services. The bank was started to service an infant who had necrotizing enterocolitis and was unable to tolerate formula, and then it expanded, as the chart in Table 2.3 presented to the Irish Perinatal Society in 2007 indicates (McCrea 2007). Cross-border cooperation for the care of Irish premature infants meant that the Irish human milk service quickly became one of the largest in the UK, as is evidenced by the following chart which was presented in 2007 to the Irish Perinatal by Ann McCrea, community health nurse by training and the original manager of this service.

As this useful chart indicates, by 2005, the Irvinestown (Sperrin Lakeland) milk bank had progressed to become the largest across the UK system (McCrea 2007, 31). In 2015, approximately 1,500 litres of milk was issued to units all over the island of Ireland, "helping approximately 856 babies, including 90 set of twins and 17 sets of triplets".[9] Although, as we will discuss, the Irish milk bank is no longer the largest milk service in the UK, it continues to be one of the largest.

Table 2.3 Volume of milk collected by milk banks (litres)

	2000	2001	2002	2003	2004	2005
Birmingham (Women's)	432	418	724	427	274	500
Cambridge (Rosie)	–	–	–	–	81	250
Chester (Countess of)	Not open as a milk bank			250	420	400
Chertsy (St Peter's)	44	25	41	25	52	44
Gillingham (Medway M'time)	35	28	40	31	20	41
Glasgow (Queen Mother's)	171	158	178	177	113	144
Huddersfield (Royal Infirmary)	105	125	187	131	135	109
Irvinestown (Sperrin Lakeland)	73	168	396	739	868	897
Kingston	91	103	107	95	21	66
London (King's College)	68	42	63	113	126	142
London (Queen Charlotte's)	216	296	323	557	437	458
London (St George's)	55	89	43	98	117	73
London (Guys & St Thomas')	Not open as a milk bank				15	190
Orpington (Princess Royal)	77	157	125	90	216	160
Oxford (John Radcliffe)	369	347	253	410	407	415
Southampton (Princess Anne)	428	524	580	513	516	600
Wirral	Not open		27	120	330	379
TOTAL	**2,164**	**2,480**	**3,087**	**3,776**	**4,148**	**4,868**

Source: McCrea 2007, 31.

Circumstances are continuing to change in the Irish human milk services, particularly in light of the UK planning to leave the EU, the so-called Brexit, which Northern Ireland and Scotland (as did London) voted against. As we discuss in Chapter 6, there have been concerns expressed regarding the Ireland wide service if and when the UK leaves the EU. Many hope and are planning for the service to continue as an island-wide health cooperation. Unfortunately, as we will discuss later in more detail, some have taken this as an opportunity to suggest that an alternative service should be opened in the southern part of Ireland. In addition, in January 2018, the milk bank moved to a larger premise at the South West Acute Hospital, meaning it is technically no longer a community-based milk bank service. Staff saw this move as a welcome change, as it meant larger and more modern facilities and access to other services, including mail, etc., and they assured me that the service would continue in the same ways, regardless of being situated in a hospital setting. However, there may be implications for moving from the community to a hospital setting which are yet to be seen.

The generosity of mothers

In 2007, the tenth anniversary conference of UKAMB was held in Chester, the current home of the largest service in the UK, the Northwest Human Milk Bank, critically discussing some of the most up to date research at the time from around the world (Wilson-Clay 2006; Vohr et al. 2006; Cohen 2007; Hartman et al. 2007; Schanler 2007). It was the first time I presented research on donor human milk services. A former donor designed the UKAMB logo (Figure 2.12) with the saying "Every Drop Counts" (Balmer 2010, 46), as is evidenced by the following screen shot from one of UKAMB's first web pages in 2002, which is archived online.

Figure 2.12 First Image for the United Kingdom Association for Milk Banking (UKAMB). Reprinted with permission from the United Kingdom Association for Milk Banking (UKAMB).

In 2014, the Northwest Human Milk Bank formed following the amalgamation of two hospital-based services, the Countess of Chester Milk Bank and the Wirral University Teaching Hospital Service, both of which relocated to the University of Chester. The Countess of Chester Hospital Service was originally called the Chester and North Wales Human Milk Bank, and was one of the only services to originally be funded exclusively from charity fundraising, which originally took three years; it was officially opened by the Duchess of Westminster on 27 June 2003. In 2000, a mother with an excess of milk asked where she might be able to send her milk. Staff arranged to have it sent to Huddersfield Milk Bank Service but decided that perhaps a more local service was warranted (Balmer 2010). The Wirral Service began in 2004 and was originally called the Wirral Mothers Milk Bank, like the Chester Service, the original idea occurred much earlier and was linked to a mother, but in the case of Wirral, the mother was bereaved after the death of her 3-week-old preterm infant. The bereaved mother than asked staff about what could be done with her stored expressed milk, and again, Huddersfield milk service was contacted. Two years later, following a UKAMB roadshow in 2001, funding for necessary equipment and other arrangements were begun (Wight 2001). But as both banks expanded, we were told, they soon realized that they were duplicating services and that a more efficient solution would be to work together and offer a larger regional service, and in turn, this meant moving to a larger premise and opportunity. The move to a community service has resulted in a much larger service, with several depots across the UK. But in some ways, this has resulted in the donor human milk services becoming more like the earlier century's discussion of "therapeutic merchandize", although the generosity of the mothers who donate is still considered the key ingredient.

"Not a Mere Metaphor"

It is widely recognized that the first to use the term "bank" for body parts was by Dr Bernard Fantus, the Chicago Cook County Hospital director of therapeutics in 1937 and originally was linked to blood, which he said was "not a mere metaphor", but that this service necessitated deposits in order to make a withdrawal, encouraging hospital staff to deposit, as well as to solicit deposits from hospital visitors (Fantus 1937, 128; Swanson 2014, 5). It is not coincidental, although much less widely known, that the first use of the term "bank" in relationship to human (mother's or breast) milk is also linked to Chicago and occurs in print for the first time four years later, in 1941, when Mary Watson, the obstetrical nursing supervisor at the Presbyterian Hospital in Chicago, published an article under the title "Our Frozen Milk Bank" (Watson 1941). Watson describes how previous methods of freezing human milk presented "too complicated and elaborate procedures" (Watson 1941, 672). Watson describes how staff collected and cooled excess milk from mothers in the hospital, which was then "boiled" and put into sterile glass bottles and then frozen and banked for future use. The process of freezing so as to preserve human milk had been reported by Emerson and Platt in 1933 from Boston, but they had not adopted the terminology of a milk bank yet. The processing (pasteurizing and freezing) of the milk was key to the origins of human milk banks, but as we have discussed, the use of the term "bank" itself was not to become widely used until much later.

We have determined that in the UK, the first time we see the term "milk bank" being used is also in 1941, and it appears in the *Derby Evening Telegraph* on 23 September 1941, but the article and reference is to a service in New York, not the UK. As the editorial states,

> An excellent medical idea which I am told has been adopted in Sydney is a milk bank in a large hospital.
>
> It is said that this system should save the lives of many babies whose mothers are unable to feed them. I cannot help thinking that it might be used in Britain's largest towns to save babies who have been left motherless as a result of raids.
>
> There is a very large milk bank in New York, and there are banks all over the United States.

As we mentioned earlier, the New York service was officially called a bureau not a bank originally, which as we also discussed is the same for the first service in UK at Queen Charlotte's Hospital, which was also originally referred to as a bureau.

The first time the term "bank" is used in the press was in the *British Medical Journal* on 27 February 1943 in a discussion about the care of the newborn and breastfeeding in a report from the British Paediatric Association subcommittee on neonatal mortality, which says, "Urges that breast-milk banks or 'pools' should be set up in maternity hospitals or other suitable centres" (BMJ 1943, 259). In March of that same year, the entire report of the British Paediatric Association is published in the *Archives of Disease in Childhood*, which specifically says,

> Breast milk banks or pools, set up in maternity hospitals or other suitable centres, would be of great value to the surrounding district and should not be difficult to maintain: the blood banks started in recent years under the stimulus of war conditions have shown what can be done by concerted and vigorous effort.
>
> (ADC 1943, 56)

This article says that this report was based on a meeting of the association which took place in London in December 1942, just after a year when it appeared to be used to discuss the Chicago milk bank service. And in the *Belfast News* on Friday 28 May 1943, there is a brief mention of a display with a

> Ministries of Food, Information and Home Affairs, and Belfast hospitals and other child welfare organisations, will have displays, and the photographic section will include pictures contributed by the Queen Charlotte's Hospital Milk Bank, in which human milk is frozen and stored for emergency distribution to babies.
>
> (*Belfast News* 1943, 3)

But the term does not seem to become used again until after the war in 1946 when a Marth Dynski-Klein, a paediatrician affiliated with two units in the west London area, including Queen Mary's, which today is affiliated with Imperial

College and Queen Charlotte's, and says that "breast milk bureaux have been created in many places", going on to say that the milk is collected

> chiefly from paid donors, living sometimes in widely dispersed areas, and require, therefore, rather elaborate arrangements for the collection, control, and preservation of human milk and the medical and nursing supervision of the donors. They are hampered by the inconstancy of supply, which depends very much on the good will of the donors and a constant and costly propaganda.
>
> (Dynski-Klein 1946, 258)

A year later, we see the first reference to "milk bank" in the *Lancet* discussing a "new" service which we are told started in 1944 at the University College Hospital Medical School and Obstetric Hospital on Huntley Street, London (which closed as an obstetric hospital in 2008), although we are not told if the term "bank" was used in 1944 or not.

There seems, therefore, to have been several services in London, and the point about "propaganda" is made again two years later by Edith Dare (1948) in her discussion of the service at Queen Charlotte's, and although she does not quote Dynski-Klein (1946) regarding the use of this term, she does quote her, saying, "Incidence of infection in infancy is estimated to be twice as high in artificially fed babies as in breast-fed ones, and the death rate ten times higher" (Dare 1948, 439; Dynski-Klein 1946, 258). It is also interesting to note that Dare uses both the term "bank" and "bureau" in the same article (a point I was told still existed into the 1990s at Queen Charlotte until at least the 1990s). Dare details the Queen Charlotte's service, which she says began in 1938. At the beginning of the article, the editors tell us that Dare recently retired after having been at Queen Charlotte's since 1911 and being made honorary director for life (Dare, 1948). An article that same year by West (1948) also discusses the tenth anniversary of the Queen Charlotte's "bank" and specifically links it to the blood services from World War II, another point which Dynski-Klein (1946) also makes. West (1948) also discusses the service which had recently opened in Cardiff, with plans for others to open in other cities across the UK. It is an important point that maternity and child welfare was a branch of the NHS which was to come into existence in July 1949 (Webster 1998). Dare (1948) also says that the service at Queen Charlotte's was originally funded by a generosity charity donation linked to the Birthday Trust and Sir Julian Cahn, but "has succeeded in becoming entirely self-supporting; indeed it is paying its own way with a reasonably safe margin" (439).

In 1944, the Welsh newspaper the Western Mail (17 May:4) reports that the city will soon be getting a "milk bank". Several years later, Greenwood Wilson (1951) reported that Edith Dare did visit Cardiff in 1944 and explained that "[d]espite war-time and post-war difficulties of materials, licensing, and controls, the Cardiff City Council at last succeeded in opening in 1947 the first municipal human milk bank in this country" (Greenwood Wilson 1951, 452). Despite this first usage of the term "bank" it was called the Human Milk Bureau still in 1983 and was located at the St David's Hospital in Cardiff (Ford et al. 1983). Despite the research links that the service at St David's Hospital had, it, like so many other services, closed in the 1980s, although it seems this closure may not have been directly related to the global problem of potential contamination by HIV/AIDs (Lucas 1987; Lucas and Cole 1990). Today, there is still not service in Wales, although, as we discussed units throughout Wales have provisions from the largest services based in England. As we discussed, the service in Ireland is also under threat, primarily due to the political circumstances beyond the scope of the service. Whereas the Scottish wide service is being supported and integrated into a government health program, and as we will discuss more later, it is being emulated around the world.

Notes

1 UKAMB was established in 1997 and states on the UK Charity Commission site that it "supports the promotion and use of milk screened according to Nice Clinical Guidance number 93. The aim of the charity is the formation of a National Screened Donor Breastmilk service that would supply infants throughout the UK according to need, not geographical location. The charity aims to support milk banks throughout the UK, by training and information sharing" (see charitycommission.gov.uk).
2 Tanya gave her first presentation to a milk banking community at the UKAMB tenth anniversary celebrations and more recently was part of UKAMB's twentieth-anniversary celebration discussing the MUIMME dissemination projects, including this book.
3 This conference featured an open talk by Mrs Margit Helleparth, the then manager of the milk banking services in Vienna. She had originally told Gillian Weaver (2007) that the bank in Vienna was established by Marie Elise Kayser, an important female German paediatrician, who is accredited with starting the first human milk service in Magdeburger, Germany, in 1919. Kayser was a breastfeeding mother herself, and it seems that after the birth of her own three children, she established a service in 1919, which did not utilize women who were working as wet nurses alone, but others who wished to help others through the collection of human milk for those in need (Volker 2015).
4 This image was reprinted with permission in a later discussion by Weaver and Williams (1997) but does not mention Calder's original article or the *Daily Herald* or Malindine.
5 These infants were reported to have been born at number 13 Ferrars Avenue, Eynesbury, Huntingdonshire, and were moved to their physician's home in St Neot's shortly after birth to accommodate the nursing staff needed for their care. Their celebrity status was to label them the St Neot's Quads.
6 A copy of this image was obtained from www.britishnewspaperarchive.co.uk, and is reprinted here with copyright permission from Independent Newspapers.
7 David Stifter, professor of Old and Middle Irish and head of the Department of Early Irish, MU, confirmed this use of the term MUIMME.
8 See www.focloir.ie.
9 See www.westerntrust.hscni.net/2026.htm.

References

Acheson, D. 1988. "HIV Infection, Breastfeeding, and Human Milk Banking." *Lancet* 2 (8605), July 30: 278.

ADC (Archives of Diseases in Children). 1943. *British Pediatric Association*. ADC, BMJ. 1 March: 55–58.

Adhisivam, B., B. Vishnu Bhat, N. Banupriya, Rachel Poorna, Nishad Plakkal, and C. Palanivel. 2017. "Impact of Human Milk Banking on Neonatal Mortality, Necrotizing Enterocolitis, and Exclusive Breastfeeding – Experience from a Tertiary Care Teaching Hospital, South India." *The Journal of Maternal-Fetal & Neonatal Medicine*, 32(6): 902–5.

AL-Naqeeb, Niran A. 2000. "The Introduction of Breast Milk Donation in a Muslim Country." *Journal of Human Lactation* 16 (4): 346–50.
Anonymous. 1935. *The Lancashire Evening Post*, 29 November, p. 5.
———. 1966. The Connaught Sentinel, p. 1.
———. 1988a. "HIV Infection, Breastfeeding, and Human Milk Banking." *Lancet* 2 (8608), August 20: 452–53.
———. 1988b. "HIV Infection, Breastfeeding, and Human Milk Banking." *Lancet* 2 (8603), July 16: 143–44. Review.
———. 1902. "Discussion." *JAMA*, 252–55.
———. 1926. "Vienna as a Post-Graduate Centre." *Canadian Medical Association Journal* 16 (2), February: 199–200.
Appadurai, Arjun. 1986. *The Social Life of Things: Commodities in Cultural Perspective*. Cambridge: Cambridge University Press.
Apple, Rima D. 1987. *Mothers and Medicine: A Social History of Infant Feeding, 1890–1950*. Madison: University of Wisconsin Press.
Arthi, Vellore, and Eric Schneider. 2017. "Infant Feeding and Cohort Health: Evidence from the London Foundling Hospital," CEPR Discussion Papers 12165, C.E.P.R. Discussion Papers.
Ascroft, Eileen. 1939. "I came to London to see the Queen." *The Daily Mirror*. Published: Thursday 30 March, p. 11.
Balmer, Sue. 2010. Milk Banking – Back to the Future. Birmingham: United Kingdom Association for Milk Banking.
Balmer, S. E., and B. A. Wharton. 1992. "Human milk banking at Sorrento Maternity Hospital, Birmingham." *Archives of Disease in Childhood*, Vol. 67(4): 556–59.
Belfast News. 1943. "Babies and Mothers 'Better Health' Campaign Belfast Programme." Friday 28 May, p. 3.
Berton, Pierre. 1977. *The Dionne Years: A Thirties Melodrama*. Toronto: McClelland and Stewart.
Boyd, C. A., M. A. Quigley, and P. Brocklehurst. 2006. "Donor Breast Milk versus Infant Formula for Preterm Infants: A Systematic Review and Meta-Analysis." *Archives of Disease in Childhood. Fetal and Neonatal Edition*, published online April 5.
BMJ (*British Medical Journal*). 1894. "Letters, Notes, and Answers to Correspondents." *British Medical Journal* 1: 1063, Published 12 May.
———. 1939. "A Human Milk Bureau Service at Queen Charlotte's," 24 June: 1298–99.
———. 1943. "ANNOTATIONS." *British Medical Journal* 1 (4286): 258 LP-260.
Budin, Pierre. 1907. *The Nursling: The Feeding and Hygiene of Premature & Full-Term Infants*. London: The Caxton Publishing Company.
Butte, Nancy, Mardia Lopez-Alarcon, and Cutberto Garza. 2002. *Nutrient Adequacy of Exclusive Breastfeeding for the Term Infant During the First Six Months of Life*. Geneva: World Health Organization.
Calder, Ritchie. 1939. "'Foster-Mothers' for the Nations Babies." *The Daily Herald*. London. Wednesday 1 March, p. 20.
Cassidy, Tanya M. 2015. 'Ireland, Irish Women and Lactation Surrogacy: Imagining a world where it takes a community to feed a child.' In Cassidy, Tanya and El Tom, Abdullahi (eds.) *Ethnographies of Breastfeeding: Cultural contexts and confrontations*. London: Bloomsbury Press, pp. 45–58.
———. 2013. 'HIV/AIDS and Human Milk Banking: Controversy, Complexity and Culture Around a Global Social Problem.' In Cassidy, Tanya M. (ed.). *Breastfeeding: Global practices, challenges, maternal and infant health outcomes*. Nova Science Publishers, Inc., pp. 93–106.

———. 2009. "Saving Little Tiny Cute Babies, Globalisation and the History of Human Milk Banking Through the Twentieth Century: A Hundred Years of Donor Human Milk Banking." *100 Years of Human Milk Banking: Looking to the Future, Learning from the Past*. Conference Proceedings. Wien, Austria.

Clarke, Joseph. 1790. "Observations on the Properties Commonly Attributed by Medical Writers to Human Milk, on the Changes It Undergoes in Digestion, and the Diseases Supposed to Originate from this Source in Infancy." *The London Medical Journal* 11 (Pt. 1): 71–91.

Cohen, R. S. 2007. "Current Issues in Human Milk Banking." *NeoReviews* 8 (7), July 1: e289–e295. American Academy of Pediatrics.

Crowther, S. M., L. A. Reynolds, and E. M. Tansey, eds. 2009. "The Resurgence of Breastfeeding, 1975–2000 The Transcript of a Witness Seminar held by the Wellcome Trust Centre for the History of Medicine at UCL, London," April 24, 2007. London: Wellcome Trust.

Daily Herald. 1939. "Life-Saving Bureau for Babies." Monday January 2, p. 3.

Dare, Edith G. 1948. "The Human Milk Bank." *Nature* 162 (4116): 439–40.

DeMarchis, A., K. Israel-Ballard, K. A., Mansen, and C. Engmann. 2016. "Establishing an integrated human milk banking approach to strengthen newborn care." Journal of Perinatology: Official *Journal of the California Perinatal Association*, 37(5), 469–74.

Department of Health and Social Security. 1981. *The Collection and Storage of Human Milk*. London: HMSO. (Report on Health and Social Subjects. No. 22).

de Meneses, Xavier, Tatiana Mota, Couto de Oliveira, Maria Inês, and Siqueira Boccolini Cristiano. 2017. "Prevalence and Factors Associated with Breast Milk Donation in Banks That Receive Human Milk in Primary Health Care Units.'" *Jorunal de Pediatria* 93 (4): 382–88.

Dunn, T. D. T., M. L. Newell, A. E. Ades, and C. S. Peckham. 1992. "Risk of Human Immunodeficiency Virus Type 1 Transmission Through Breastfeeding." *Lancet* 340: 585–88.

Dynski-Klein, M. 1946. "Breast Milk Bank in Maternity Units." *The British Medical Journal* 2 (4468): 258–60. www.jstor.org.jproxy.nuim.ie/stable/20367235

Fantus, B. 1937. The therapy of the Cook County Hospital July 10. *Journal of the American Medical Association*, reprinted 1984; 251: 647–49.

Fildes, Valerie. 1986. *Breasts Bottles and Babies*. Edinburgh: University Press Edinburgh.

———. 1988. *Wet Nursing a History from Antiquity to the Present*. Oxford: Basil Blackwell.

Ford, F. 1949. "Feeding of Premature Babies." *Lancet* 253 (6563), June: 987–94.

Ford, J. E., Alicja Zechalko, J. Murphy, and O. G. Brooke. 1983. "Comparison of the B vitamin composition of milk from mothers of preterm and term babies." *Archives of Disease in Childhood*, 58: 367–72.

Greenwood Wilson, J. 1951. "Random Reflections on a Human." *Archives of Disease in Childhood* 26: 452–56.

Grøvslien, A., H. H. Torng, G. E. Moro, J. Simpson, and D. Barnett. 2013. International Perspectives on Donor Milk in and beyond the NICU. *Journal of Human Lactation*, 29(3): 310–12.

Golden, Janet Lynne. 1988. "From Wet Nurse Directory to Milk Bank: The Delivery of Human Milk in Boston, 1909–1927." *Bulletin of the History of Medicine* 62 (4), Winter: 589–605.

———. 1996. *A Social History of Wet Nursing in America: From Breast to Bottle*. Cambridge: Cambridge University Press.

Hartmann, B. T., W. W. Pang, A. D. Keil, P. E. Hartmann, and K. Simmer. 2007. "Best Practice Guidelines for the Operation of a Donor Human Milk Bank in an Australian NICU." *Early Human Development* 83 (10), October: 667–73.

Hewlett, B. S., and S. Winn. 2014. "Allomaternal Nursing in Humans." *Current Anthropology* 55: 200–29.

Holt, L. E. 1901. *The Care and Feeding of Children.* New York, NY: Appleton.
Holt, L. E. 1957. *The Good Housekeeping Book of Baby and Child Care.* New York, NY: Popular Library, Inc.
HIV in Zimbabwe: A Qualitative Study." *Journal of Human Lactation* 22 (1): 48–60.
Irish Business and Employers' Confederation. 2007. *Business Perspectives on Future Dairy Policy: A Discussion Document Compiled by the Irish Dairy Industries Association.* Dublin: Irish Business and Employers Confederation, Dublin.
Jones, Frances. 2003. "History of North American Donor Milk Banking: One Hundred Years of Progress." *Journal of Human Lactation* 19 (3), August: 313–18.
Kepler, Paul, mit einen Beitrag and von Helmut Gadner. 1988. *Das Kind und sein Arzt; 150 Jahre St. Anna Kinderspital.* Wien: Facultas-Universitätsverlag. www.stanna.at/geschichte
Kiriline, Louise. 1936. *The Quintuplets' First Year; The Survival of the Famous FIVE DIONNE BABIES and Its Significance for All Mothers.* Toronto: The Macmillan Company of Canada Limited.
Kirkpatrick, T., C. Percy, and Henry Jellett. 1913. *The Book of the Rotunda Hospital: An Illustrated History of the Dublin Lying-in Hospital from Its Foundation in 1745 to the Present Time.* London: Adlard & Son, Bartholomew Press.
Koch, J. T. 2006. *Celtic Culture: A Historical Encyclopedia Volumes 1–5.* California: ABC-CLIO Inc.
Lancet. 1938. "A Bureau for Human Milk." *Lancet* 232 (6014), December: 1307.
Lucas A. 1987. "AIDS and Human Milk Bank Closures." *Lancet* 1 (8541), May 9: 1092–93.
Lucas, A., and T. J. Cole. 1990. "Breast Milk and Neonatal Necrotising Enterocolitis." *Lancet* 336 (8730), December 22–29: 1519–23.
Marx, Karl. 1977. *Capital: A Critique of Political Economy,* Vol. i, Translated by Ben Fowkes. New York: Vintage Books.
Mayerhofer, E. u. Přibram, E. 1909a. "Ernährungsversuche mit konservierter Frauenmilch." Wiener klin. Wochenschr. Nr. 26. (Feeding Experiments with Conserved Woman's Milk).
———. 1909bM. "Über Ernährung mit konservierter Frauenmilch." Verhandl. d. 26. Vers. d. Gesellseh. f. Kinderheilk. in Salzburg (1909) 99. (On Food Including Canned (or Conserved) Woman's Milk).
Mayerhofer, E., and Přibram, E. 1912. "Praktische Erfolge der Ernährung mit konservierter Frauenmilch (Bericht über 100 Fälle)." *Zeitschrift für Kinderheilkunde* 3 (1): 525–67.
McCrea, Ann. 2007. "The Human Milk Bank – Cross Border Co-Operation Looking After the Irish Premature Baby." Abstract from the Irish Perinatal Society. archive.imj.ie
National Library of Canada. 2003. "*Dionne Quintuplets Digitization Project.*" Accessed September 20 2003. www.nlc-bnc.ca/initiatives-bin/rella?mode=nbr&project_nbr=4915
O'Reilly, E., and J. O'Donovan. 1817. *Sanas Gaoidhilge-Sagsbhearla, An Irish–English Dictionary . . . to which is annexed, a compendious Irish Grammar.* (There are several versions and the 1821 version is available on Google Books.)
Paterson, Donald. 1938. "Note to Edith Dare." National Trust Birthday Fund (NTBF) Archives, Wellcome Trust.
Paterson, Donald, and J. Forest Smith. 1939. *Modern Methods of Feeding in Infancy and Childhood.* 7th ed. London: Constable.
Porter, Roy. 1987. "A Touch of Danger: The Man-Midwife as Sexual Predator." In *Sexual Underworlds of the Enlightenment,* edited by G. S. Rousseau and Roy Porter. Manchester: Manchester University Press.
Rosack, M. L. 2004. "The Dionne Quintuplets Perinatal Care a la 1930s Style." *Awhonn Lifelines* 8 (4): 348–55.
Schanler, R. J. 2007. "Evaluation of the Evidence to Support Current Recommendations to Meet the Needs of Premature Infants: The Role of Human Milk." *The American Journal of Clinical Nutrition,* 85 (2), February: 625S–628S.

Schuman, Andrew. 2003. "A Concise History of Infant Formula (Twists and Turns Included)" (HTML). Contemporary Pediatric, February 01. Accessed September 16 2006.

Seifert, Rita. 2012. "Erste promovierte Ärztin der Universität Jena und Begründerin der Frauenmilchsammelstellen in Deutschland (Marie-Elise Kayser (1885–1950) First doctorate the University of Jena and founder of the women's milk collecting points in Germany)." *Weimar – Jena: Die große Stadt* 5 (2): 111–20. www.verlagvopelius.eu

Silverman, William A. 1979. "Incubator-Baby Side Shows." *Pediatrics* 64 (2): 127–41. Reproduced by Permission of *Pediatrics*. www.neonatology.org/classics/silverman/silverman1.html

Slimes, Martti A., and Niilo Hallman. 1979. "A Perspective on Human Milk Banking, 1978." *Journal of Pediactrics* 94 (1), January: 173–74.

Solomons, B. 1934. "Report of the Rotunda Hospital." *Irish Journal of Medical Science* 6 (104): 331–86.

Sussman, George. 1982. *Selling Mother's Milk: The Wet-Nursing Business in France.* Urbana: University of Illinois Press.

Swanson, Kara W. 2014. *Banking on the Body: The Market in Blood, Milk, and Sperm in Modern America.* Cambridge, MA: Harvard University Press.

Titmus, Richard. 1997 [1970]. *The Gift Relationship: From Human Blood to Social Policy.* Reprinted by the New Press. Reissued with new chapters, John Ashton, and Ann Oakley. London: LSE Books.

Tobey, James A. 1929. "A New Foster-Mother." *Hygeia* 7: 1110–12.

Vohr, B. R., B. B. Poindexter, and A. M. Dusick, et al. 2006. "Beneficial Effects of Breast Milk in the Neonatal Intensive Care Unit on the Developmental Outcome of Extremely Low Birth Weight Infants at 18 months of Age." *Pediatrics* 118: e115–e123. http://pediatrics.aappublications.org/cgi/content/full/118/

Volker Klimpel. 2015. "Marie-Elise Kayser" In *Biographical Encyclopedia for Nursing History "Who was who in nursing history",* Vol. 7, Hubert Kolling (ed.) hps media Nidda. pp. 142–43.

Watson, Mary L. 1941. "Our Frozen Milk Bank." *The American Journal of Nursing* 41 (6): 672–74.

Weaver, Gillian. 2007. Personal communication by email to Tanya Cassidy, and discussed with permission.

Weaver, G. and A. S. Williams. 1997. "A Mother's Gift: The Milk of Human Kindness." In *The Gift Relationship*, edited by Titmuss, R. M., A. Oakley, and J. Ashton. 2nd ed., 319–32. New York: The New Press.

Webster, Charles. 1998. *The National Health Service: A Political History.* Oxford: Oxford University Press.

West, Paul. 1948. "Medical News – Human Milk Banks: Novel Method of Fighting Infant Mortality." *Indian Medical Gazette* 83 (3): 143–44.

Wharton, B. A. 1981. "Immunological implications of alternatives to mother's milk. II Donormilk." In: Wilkinson, A. W., ed. *The immunology of infant feeding.* New York: Plenum Press, pp. 123–35.

WHO (World Health Organization). 2003. *Global Strategy for Infant and Young Child Feeding.* Geneva: World Health Organization.

Wight, N. E. 2001. "Donor Human Milk for Preterm Infants." *Journal of Perinatology* 21: 249–54.

Wilson-Clay, Barbara. 2006. "The Milk of Human Kindness: The Story of the Mothers Milk Bank at Austin." *International Breastfeeding Journal* 1: 6. Published Online March 27 2006. doi:10.1186/1746-4358-1-6

3 Building the science and society of human milk with banks

From the very beginning of donor human milk banking services, as discussed in the previous chapter, the presentation of science and technology has been all pervasive. When you visit a donor human milk service bank for the first time, all of the technology involved is on display, including pasteurizers, human milk analysers, laminar or biosafety airflow cabinets, as well as less obviously scientific technology, such as freezers, fridges, dishwashers, computers, telephones and other office equipment, all of which originally were once state-of-the-art forms of technology, but which have not been changed as these items become more taken for granted. In this chapter, we wish to explore the science and technology surrounding the social and cultural contexts of donor human milk services in the UK. One of the staff told me early in my fieldwork that donor human milk services is not "rocket science", but as we will see, science and technology inform every aspect of these services, and some of the current science of human milk has some highly important scientific implications for future healthcare provisions and can in some cases literally be considered a matter of life and death. Donor human milk services integrate science and technology throughout the process, including the technical processing and testing involved with the milk, although there are rapid changes occurring around the world at the moment (Moro et al. 2019). The donors are screened through various blood tests, and many of the recipients have experienced highly medicalized births, which means they are surrounded by neonatal technologies in NICUs in hospital settings. In this chapter, we will begin with some of the key scientific research questions associated with donor human milk. The earliest questions were often linked to comparative composition with other mammals and, ultimately, were connected with the marketing of artificial forms of infant "formula", which were originally based on clinically informed experiments about how to modify alternative animal-based milk to more closely resemble human milk, adverts which continue today when the $70 billion infant formula industry evokes their more than century-old research into the science of human milk. The science of human milk can also, as we will discuss, be used to expand practices with regard to human milk services globally, but as we will continue to discuss in Chapter 6, these expansions must always be aware of potential exploitation.

Questioning human milk science

As we discussed in the previous chapter, some of the key early questions associated with human milk science were related to the preservation of human milk after it has been removed from a mother's body, an issue which is particularly important when we are discussing feeding vulnerable infants who may not have the ability to go to breast. Also, we must recall that donor human milk services began in an age when not only was refrigeration less common but also freezing essentially unknown. Therefore, the discovery of preservation methods represented a decisive innovation, and the expansion of pasteurization became a key issue linked not only to preservation but also, as Bruno Latour has discussed extensively, to the science of bacteriology. Peter Atkins (2000, 2010, 2016) has extensively discussed pasteurization in Britain in relationship to bovine milk, quoting from a *Dictionary of Dairying* published in 1950, which remarked, "Probably no subject outside of religion and politics has been the cause of more prolonged and bitter controversies than the proposal of compulsory pasteurization of all milk" (Atkins 2000, 41). Atkins goes on to point out that the adoption of pasteurization was slow in Britain. Although the first commercial equipment was available in Germany from the 1880s, in Copenhagen and Stockholm, milk was routinely pasteurized by 1885; meanwhile, in Britain, by 1926, only 1.5 percent of British supplies were pasteurized, with the majority of milk still being raw until after the Second World War (Atkins 2000).

In 1984, Bruno Latour published *Les microbes : guerre et paix ; suivi de, irréductions* (Latour 1984), which was published in English three years later by Harvard under the title *The Pasteurization of France*.

> To understand simultaneously science and society, we have to describe war and peace in a different way, without ourselves waging another war or believing once again that science offers a miraculous peace of mind.
>
> (Latour 1988, 6)

In one of the first studies to extensively integrate the world of science with society, Latour argues that the events surround the pasteurian experiments integrated pasteurians with hygienist creating bacteriology. He goes on to say, "The pasteurization of beer or milk . . . were only demonstrative and efficacious, only in the laboratory", and for these applications to spread, settings, such as hospitals, needed their own laboratories (Latour 1988, 90).

> If Pasteur had written a work on the sociology of the sciences, he might have entitled it "Give me a laboratory and I shall raise the world".
>
> (Latour 1988, 90)

A relevant circumstance that is not discussed by Latour, is the fact that France has one of the largest human milk systems in Europe (see EMBA), and I visited the regional service in Lyon during our EU research project. In France, the majority of milk which is processed in the human milk bank is MOM, and since all milk used for infants in the NICUs is pasteurized, only a small percentage of the human milk processed is donor human milk.

The milk laboratory became key not only to the science of human milk but also the clinical human milk services. As we discussed in the previous chapter, the pasteurization of human milk was considered by many to be a key feature associated with the origins of donor human milk services (Mayerhofer and Přibram 1909a, 1909b). Central to the birth and expansion of the medicalized use of human milk are the early studies that came out of Vienna, arguably the most important research in paediatric microbiology, under the guidance of the "first pediatric infectious diseases physician", Theodor Escherich (Shulman et al. 2007, 1025), who is accredited by the junior researchers in writing with having the idea to pasteurize human milk. After intensive laboratory experiments, Escherich (1857–1911) published his work on intestinal bacteria and the infant gut, describing what he originally called "bacterium coli commune", but which was to posthumously be named Escherichia coli or E-coli for short and was to solidify him as one of the leading bacteriologists in paediatrics. In 1890, Escherich moved to St Anna Children's Hospital in Graz, and then in 1902, he moved to Vienna as a full professor of paediatrics at the University of Vienna and the St Anna Children's Hospital, the most prestigious paediatric post in Europe. Escherich himself had long studied the gut microbes of infants fed breastmilk and was a campaigner for breastfeeding for many years, being a key link between the preservation of this precious bodily fluid and the scientific purification of this "therapeutic merchandize".

MOM and therapeutic merchandize (or "To Heat or Not To Heat")

MOM is always the best choice for infants, since the milk, as we discuss more next, has specific bioactive components linked to the complex maternal-infant dyad. The units involved in our EU study, and most units across the UK, do not normally pasteurize MOM. However, as a Belgian randomized control trial on whether to pasteurize or not to pasteurize MOM states, "Due to lack of microbiological standards, practices such as pasteurization of mother's own milk differ widely among neonatal intensive care units worldwide" (Cossey et al. 2013, 170). The Lyon regional milk banking team recently published a discussion saying that MOM is pasteurized, particularly in France, most "notably when the mother is positive for cytomegalovirus (CMV), when the collection has not been performed in good hygienic conditions, or when the milk has been stored for more than 48 to 96 hours" (Picaud and Buffin 2017, 99), showing the following image which represents the decision making process (Figure 3.1).

As part of our EU-funded research, we were able to visit this service in Lyon, a research expedition particularly relevant in light of the fact that the head of EMBA is the clinical lead for this donor human milk bank. In the UK, during our ethnographic research, only one of the milk banks was asked by a hospital to pasteurize MOM, which was an unusual event and was due to a potential failure in a local NICU freezer system, but as we have already observed, this procedure is not the normal operation for UK based milk banking services.

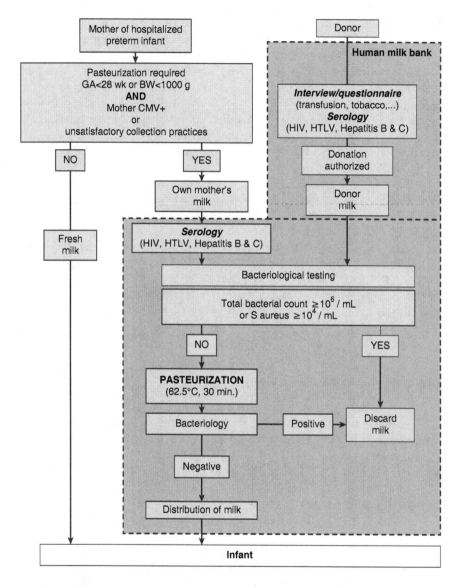

Figure 3.1 Human milk banks and feeding preterm infant with human milk (HM), birth weight (BW), gestational age (GA).
Source: *Clinics in Perinatology.* 2017, 44(1). Picaud and Buffin, "Human milk: treatment and quality of banked human milk."

There are different kinds of pasteurization; for instance, some kinds do not constitute sterilization, as sterilization kills too many of important oligosaccharides. Human milk banks services, as we will discuss more in the next chapter, consider the pasteurization of milk a process, and like other operating procedures of the donor human milk services in the UK, it is guided by the National Institute for Health and Clinical Excellence (NICE) guidelines (2010) titled "Donor Breast Milk Banks: The Operation of Donor Milk Banks Services". These guidelines provide evidence-based details about running this service, and because it is linked to the NHS, there are also procedures for governance and compliance. As the NICE (2010) guidelines state,

> There are many methods of treating milk with heat, usually with the aim of pasteurisation. As with the storage of milk, the balance of benefits from raw or minimally treated milk, with harm from contaminated or heavily processed milk, need to be considered.
>
> (NICE 2010, 90)

The same document then goes on to say that most banks use *low-temperature, long-time pasteurization*, more commonly known as the *Holder Pasteurization* (HoP) method of pasteurization, which involves raising the temperature of the milk to 62.5° C (145° F) for 30 minutes, but this method is known to impair immunoglobulins, enzymes, cytokines, growth factors, hormones or oxidative stress markers (Escuuder-Vieco et al. 2018). This was the historical method for pasteurizing human milk, but in the dairy industry, and in various parts of the world, high-temperature short-time (HTST) pasteurization is preferred because of the preservation of some components in the milk. It is agreed that more research is needed. HTST or flash pasteurization has been the subject of a lot of research recently, including some interesting recent work in Spain (Escuder-Vieco et al. 2018) and some important commercialized work from South Africa, produced in conjunction with major milk banking staff, resulting in a portable cost-effective system called PiAstra (see piastra.org).

The four milk banks involved in our research all use machines which are based on the Holder pasteurization method, but two used one type of machine made by ACE and two used one type of machine made by Sterifeed, which involves heat-sealed containers and the immersion of the containers in water which is heated and then cooled.

Bacteriology

According to the NICE guidelines, a test sample from each batch of pooled (pooled, as we remember, from only one donor) milk should undergo microbial contamination testing and be discarded if samples exceed the following counts:

- 10^5 colony-forming units (CFU)/ml for total viable microorganisms **or**
- 10^4 CFU/ml for Enterobacteriaceae **or**
- 10^4 CFU/ml for Staphylococcus aureus.

Additionally, the NICE guidelines state that similar tests should be performed regularly on pasteurized milk to determine if there is any potential microbial contamination, but this is primarily to make sure that the pasteurizing machines are working properly.

Tracking and tracing

Being able to track and trace any potential contaminants associated with donated human milk has ensured that in the UK, NICE guidelines suggest that milk should not be pooled from more than one mother, a policy which is not enforced in other countries, such as the US, where milk is pooled from more than one mother in order to increase the potential fat components. However, in the UK, it is more important to be able to trace the milk directly back to the donor, resulting in the UK system being compatible, as we will discuss in Chapter 6, with Islamic milk kinship laws. The Scottish service has been actively involved with the ICCBBA (International Council for Commonality in Blood Banking Automation) linked to the WHO, although based in California, and responsible for the ISBT (International Society of Blood Transfusion, reflecting the original important role of this group) 128, a global standard for the "identification, labelling, and information transfer" of medical "products of human origin" (originally for blood, but extended to include other tissues of human origin, including most recently human milk) across different healthcare systems, as well as across international borders (see Cabana 2016). Across the world, donor human milk is classified alternatively either as food or as tissue, or as a hybrid of both, which makes a difference in terms of regulation. ICCBBA would clearly like to see donor human milk to be classified within the same frame as blood, and other products of human origin. If classified the same as blood, there would be major implications for comparative research facilities planning to cut across borders with breastmilk studies, such as on the island of Ireland, where donor human milk is classified as food.

In addition, the ability to be able to track and trace the journey of the milk from donor, to bank and to recipient is also a key issue. Again, in Scotland, because a centralized service is used across the country, a distinctive system has been derived, although this was a hybrid model and is not being developed further. During our research, two of the services involved in our study met several times with a commercial company that was developing a system they hoped would become the UK's, and beyond, standard for tracking and tracing human milk either in milk banking services or for neonatal units with regard to the use of MOM. Li-Lac has been adopted by one service outside of the NHS in the UK but has not expanded to other services due to cost implications, although this may change in the future. The additional problem is that this is not the only system available within the UK, and there are additional systems in France which have been used for many years.

Donors

The NICE guidelines also give details regarding what should be discussed in respect to the recruitment and screening of donors. It suggests that once a donor has been identified or approaches the service, potential donors should first undergo a verbal screening, which should include the necessity to have an additional blood test, which will involve serological screening for HIV 1 and 2, hepatitis B and C, HTLV (Human T-cell lymphotropic virus) I and II, as well as for syphilis, and that antenatal tests are not sufficient to be eligible to become a donor. For some people, these tests, or the fear of needles, may be reason not to donate, but for others, this is an understandable step, which guidelines say does not need to be repeated again during the donation period, although if someone returns to be a donor after a subsequent pregnancy, the tests must be performed again. All of the milk banks send kits with letters of explanation to the potential donors who are then able to take these materials to their local healthcare providers who can then draw the samples of blood, which are then returned to the bank itself and sent out to be analysed, usually to the serological testing services affiliated with the hospitals linked to the service.

The science of hygiene frames a critical discussion for donors, as the staff wish for the milk to arrive as free from bacteriological contaminants as possible. As part of this discussion, all milk bank staff members discuss with donors the best practices for hand washing and equipment washing, and some require breast washing, although most suggest that daily routines of breast cleanliness should be sufficient.

Additionally, donors are also often confronted with the need to use both refrigeration to cool down their expressed milk and freezing to extend the length of time before the milk needs to be transferred to the bank itself. Separate equipment is not purchased, although keeping the milk in separate areas in both the refrigerator and the freezer are often requested. Some people do want to purchase separate units, but most do not.

The last technology of donation which we wish to discuss is the breast pump, described by some as a feminist technology (Boyer and Boswell-Penc 2010), but which is seen by others to represent an increasing divide between mother and baby (Thorley 2011), potentially leading to increased use of bottles (Van Esterik 1996). The breast pump is now considered, in at least some parts of the world, to be an essential part of maternal experiences (Lepore 2009), and as a recent UK study found, it is considered important to mothers as a healthcare provision (Crossland et al. 2016).

What's in the milk?

Older than the science of pasteurization (as we also mentioned in the previous chapter) are compositional studies, and although we have known for centuries about the basic components in human milk (Clarke 1790), as the anthropologist Katie Hinde said in her Ted Talk "What We Don't Know about Mother's Milk", we know more about coffee and caffeine than we do about human milk. Recently, knowledge about the composition of human milk has been expanding exponentially, with several reviews of both the nutrients and the bioactive factors available (Ballard and Morrow 2013; Andreas et al. 2015; Mosca and Giannì 2017; Dror and Allen 2018). Human milk is a dynamic bioactive fluid, changing throughout lactation and related to the dynamics of the maternal/infant dyad. Even the *Encyclopædia Britannica* has recently published on lactation with a section regarding the composition of milk (Donovan 2018). Many of these early studies were linked to comparisons with mammalian secretions from other animals, in particular bovine milk, itself a vast global industry, despite the fact that some other animals have milk closer in composition to human milk (goat, for instance; see Bosworth and Van Slyke 1916). Meanwhile, the global dairy industry is heavily invested in infant formula, and as we mentioned in the previous chapter, Ireland plays a comparatively large role in this industry on a global scale.

Table 3.1 Lactation Biology

Some constituents of human colostrum, transitional, and mature milk and of cow's milk (average values per 100 millilitres whole milk)				
	colostrum (1–5 days)	transitional (6–14 days)	mature (after 14 days)	cow's milk
energy, kcal*	58	74	71	69
total solids, g	12.8	13.6	12.4	12.7
fat, g	2.9	3.6	3.8	3.7
lactose, g	5.3	6.6	7.0	4.8
protein, g	2.7	1.6	1.2	3.3
casein, g	1.2	0.7	0.4	2.8
ash, g	0.33	0.24	0.21	0.72
Minerals				
calcium, mg	31	34	33	125
magnesium, mg	4	4	4	12
potassium, mg	74	64	55	138
sodium, mg	48	29	15	58
iron, mg	0.09	0.04	0.15	0.10

*Kilocalorie; sufficient energy to raise the temperature of 1 kilogram of water 1 degree Centigrade.

Source: Donovan 2018.

The chart illustrates only some of the constituents of this complex, dynamic, live biofluid (Table 3.1). A more complex discussion of human milk was part of our ethnographic study and is linked to the so-called breastfeeding ad produced by a group under the title "Human Milk, Tailor-Made for Tiny Humans" (Figure 3.2). This advertisement first aired in October 2016 at the annual UNICEF conference during a presentation by one of their technical advisors to the project in a talk entitled "Examining Psychological, Social and Cultural Barriers to Responsive Breastfeeding: Who Really Decides How Women Feed Their Babies and What Can We Do about It?", a presentation which argues that one of the things parents wanted to see was more discussion of science (Brown 2016). In January 2017, the full advert was official launched at the Science Museum in Bristol, and I was invited to the launch. Later, the founder and director gave an interview to Kellymom.com and described this initiative as "a collective of parents working together, mostly remotely, to share

Building the science and society of human milk 63

the science of human milk with as many people as we can" (Tchaikowski on Kellymom.com 2018). Generously, the producers of this advert have shared this commercial with breastfeeding researchers around the world, and they have allowed us to reprint their infographic here as a visual display of the science of breastmilk.

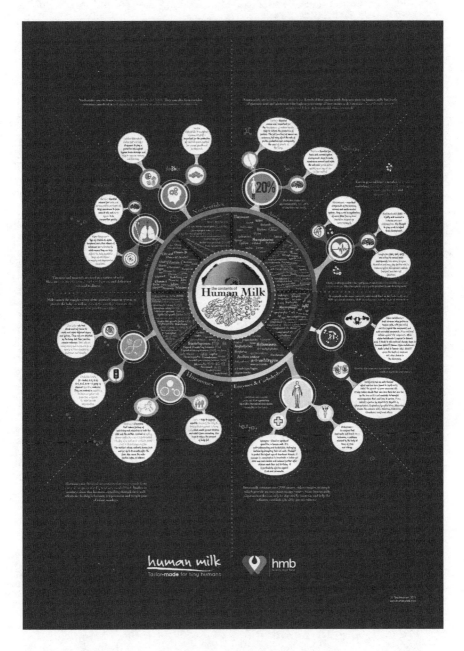

Figure 3.2 Human Milk – Tailor-made for Tiny Humans – Contents of Human Milk
Source: infographic. Reprinted with kind permission from Claire Tchaikowski, CEO of Tiny Humans CIC, Bristol, UK. www.human-milk.com

64 *Building the science and society of human milk*

This infographic was constructed in connection with a milk banking service in London, and an interactive version can be found at human-milk.com under "science". Included are the following references which were used for all of the major points, including answers to "nature has been researching your milk for hundreds of millions of years" (BBC; Oftedal 2002a, 2002b; Capuco and Akers 2009; Goldman 2002). In support of "your milk contains ingredients that kill cancerous cells", we find Gustafsson et al. 2005; Håkansson et al. 1995; Hallgren et al. 2008; Håkansson et al., 2011; Kataev, Zherelova, and Grishchenko 2016; Jiang, Du, and Lönnerdal 2014; Hill and Newburg 2015; Vogel 2012. They then list references for the statement, "Your milk contains stem cells. These are cells that create and repair the body, and are being researched worldwide to cure conditions like Alzheimer's and diabetes" (Cregan et al. 2007; Hassiotou et al. 2012; Briere et al. 2016; Twigger et al. 2015; Choi, Lee, and Lee 2016; Lilly et al. 2016; Cheng et al. 2016). The next point which they provide reference for concerns bacteria and viruses, as they state, "Your body identifies bacteria and viruses found in your baby's body and environment. You then produce antibodies specifically tailored for those infections and deliver them to your baby through your milk. The more milk she drinks, the more antibodies she receives" (Goldman et al. 1982; Pickering and Kohl 1986; Litwin, Zehr, and Insel 1990; Blais, Harrold, and Altosaar 2006; Andreas et al. 2015; Bode 2015; Hassiotou and Geddes 2015; Turfkruyer and Verhasselt 2015; Bourlieu and Michalski 2015; Rogier et al. 2014).

The references then turn to the hormone leptin, saying, "Your milk appears to switch on a gene in your baby's body, which produces a hormone called leptin. This hormone tells your baby when his tummy is full, protecting him against over eating" (Fields and Demerath 2012; Savino et al. 2016; Miralles et al. 2006; Cannon et al. 2015). The referencing list ends with sources for research on Oxytocin saying, "Your milk contains Oxytocin, a hormone that induces relaxation, and feelings of well-being in your child and in you" (Unvas-Moberg 1997; Groer and Davis 2002; Winberg 2005; Strathearn 2011; Vargas-Martínez 2017; Jonas 2016).

The next necessary question involves the composition of pasteurized donor milk. In a recent extensive review of research about components of donor human milk which has undergone HoP, an Italian team (Peila et al. 2016), which includes founding members of the EMBA, notes

> Saccharides are not significantly affected by the heat treatment, as either free molecules or as part of biologically active compounds. The total lipid content is preserved by HoP, as is its fatty acid composition. This finding is of paramount importance since it suggests that pasteurization is able to preserve both the nutritional and biological properties relevant to the development of the central nervous system associated with some of these fatty acids. Consistently, fat soluble vitamins also seem to be unaffected, while water soluble vitamins, and vitamin C in particular, are generally reported as significantly decreased. The results concerning specific biologically active molecules (such as cytokines and growth factors) remain uncertain, due to

the vast number of different compounds analyzed in each study, and to the paucity of comparable results.

Proteins are more significantly affected by HoP. In fact, specific proteins with significant immunologic and anti-infective action (such as immunoglobulins and lactoferrin) are reduced by pasteurization.

(Peila et al. 2016, 8)

This review points out several issues that are inconsistent between studies, saying that the tests themselves often do not take into consideration the dynamic variability of MOM, our favourite acronym for mother's own milk, as it also represents the agency behind this dynamic bodily fluid, an agency we will discuss in greater detail in Chapter 5. As we have mentioned, other forms of pasteurization, such as HTST, are less compromising to the milk in terms of some aspects, finding that "despite two immune components (IgA and lactoferrin) being negatively impacted, a further four components (IL-10, IL-8, lysozyme, and oligosaccharides) were unaffected or minimally affected" and that the associated negativity was less than with HoP using the Sterifeed method (Daniels et al. 2017a). In a simultaneous study of the HTST PiAstra system reports no loss regarding human milk oligosaccharides (HMO) (Daniels et al. 2017b), after lactose and fat, HMOs are the third-most "largest component" in human milk (Chen 2015) and are complex sugars increasing the considered key for an infant's gut microbiome, which are wonderfully not destroyed during pasteurization.

However, the cost of implementing HTST on a larger scale continues to be prohibitive. Similarly, the cost of using unpasteurized MOM is exorbitant, involving the need to perform tests on all patches individually, resulting in the loss of higher proportions of milk donated, and only a few places are willing and/or able to incur such costs (Grøvslien and Grønn 2009). Regardless of these drawback, as the Italian reviewers we discussed earlier mentioned at the end, their important discussions on "clinical practices demonstrate that many beneficial properties of human milk remain, even after pasteurization" (Peila et al. 2016, 8), which is a clinical practice which we will discuss more in the next chapter.

Marketing the science of breastmilk

The relationship between donor milk banking and human milk research is a complex one (Shenker 2017). Our research enables us to postulate four essential categories of milk banking services based on links between clinical and research issues. By establishing a taxonomy of human milk services, it may be possible to talk about harmonization of best practice—an outcome which will involve not only the dominance of any one particular model but also will provide a recognition of the distinctive strengths and opportunities of each particular model. Healthcare policy and governance may thereby be equipped to consider in informed terms the balance of donor human milk services within relevant jurisdictions.

The first type of bank (type a) is a clinical bank with few or no links to a research laboratory. These banks typify a certain efficient simplicity demonstrated by the quoted phrase, "It's not rocket science". Such banks are likely to stress the obviousness of human milk as a "natural" choice and the clinical need for human milk while nonetheless adhering to the most stringent and scientifically demonstrated standards of hygiene and safety. Several of the smaller milk banks would fit into this category, especially those that only service their own units, although it is not unheard of for these units to participate in research; it is just much less a priority.

A modification of this first category would involve a clinical bank in which a few bio-samples are regularly but not routinely stored for possible research. The second type of bank (type b) describes those clinical and research banks in which milk research is a normal and expected function of the bank, but the day-to-day operation of the bank is dictated by clinical concerns. A key to this definition of type b would be the proviso that only milk which cannot be used can be directed towards research. Only when clinical demands are satisfied can research be either initiated or developed. These two types, in fact, capture the features of the four main milk banks involved in our ethnographic research. All four banks are involved in research, although for both of the community-based banks, research is not as key a focus, although they do participate in it and are extremely cooperative, their main focus is clinical and the supply of DHM for clinical uses.

A third type (type c) describes a bank in which research imperatives define the organization of the bank, but which also routinely supplies donor human milk banks for clinical purposes. An interesting example of this bank would be the Prolacta bank, which focuses on human milk research for commercial purposes, but with a loudly advertised altruistic agenda. The two hospital-based large UK banks which participated in our ethnographic study saw research as key, although they also were very clear about the clinical uses, but as university hospitals, they emphasized the need for research.

The fourth or final type of bank (type d) describes a bank that is wholly devoted to research with no clinical responsibilities whatsoever. Currently, there is no bank in the UK which fits this type. However, in the US, there is a such a bank type represented by the Mommy's Milk Human Milk Biorepository at the University of California San Diego (UCSD), which seems to be linked to some very interesting and significant breastmilk scientific research.

Another complicated example can be found through an analysis of the (albeit not yet complete) online Global Biobank Directory, Tissue Banks and Biorepositories, which lists the Coreva Human Milk Bank in Westlake Village, California, as the only, explicitly stated, biobank with human or breastmilk. However, the hyper-link sends you to a web page that says the account is suspended for nationalmilkbank.org. If we look at the archives web pages for nationalmilkbank.org, we see that the National Milk Bank (NMB) was started in 2005, announced as "the nation's first virtual human milk donation organization" and linked to the commercial company Prolacta Bioscience, which dates back to 2001. Prolacta is a controversial human milk company in the US, which is linked to a number of non-HMBANA (Human Milk Banking Association of North America) endorsed milk banks, in other words, to a set of DHM banks that provide Prolacta with most of the milk which is donated, from which they produce human milk products which are then sold to hospitals. There continue to be tensions between this commercial form of DHM banking and the not-for-profit banks affiliated with HMBANA and UKAMB, uneasiness which may have been warranted when the company began, especially when we consider some potentially ethically dubious links which were original made on Prolacta's early web pages back in 2001. We can see that there are discussions of information about donor milk banking pointing the reader to two *Discover Magazine* articles that examine T. Colin Campbell's work on inter-species consumption and health (controversially later published with his son in the China Study; see Campbell and Campbell 2005), as well as Catharina Svanborg's research on breastmilk's cancer-killing properties (first published in 1995; see Håkansson et al.), thus setting the early stage for linkages between the "magic" of DHM and hard science. Interestingly, the Campbell and Campbell book has since been criticized for its extensions beyond the scope of the original data. Also, Svanborg has discussed her interpretation of why her team was not inundated with offers to extend their research, which despite these setbacks has continued to develop important connections involving human milk, although not necessarily pasteurized DHM, and cancer-killing agents.

The links between human milk science and cancer research have also been discussed by Kathleen Arcaro, who has been working with breast cells that are donated when DHM is expressed. Kathleen Arcaro and her team looked at samples of breastmilk from lactating women scheduled for breast biopsy, which provides epithelial cells which can be analysed for hyper-methylation and therefore as indicators for breast cancer risks (Murphy et al. 2016). Arcaro's work called on the so-called Army of Women, which was set up by Dr Susan Love to conduct widespread research on breast cancer and uses technology to bring together large groups of women to study breast cancer–related issues. In 2012, they helped with the so-called Milk Study (Wilson et al. 2015). The effort of recruiting these women, organizing them and communicating the value and purpose of the research itself represents a methodological intervention into the habitual terrain of social science. When scientific data must be volunteered from human subjects, then the social sciences and even the humanities can contribute to rigorous empirical research.

Public trust is essential not merely to secure funding, but also to focus human milk donation campaigns effectively. A degree of transparency is required to preserve and sustain trust. Unregulated donor human milk research which overlaps with clinical usage risks eroding public trust if resentment is provoked by the perception that milk is not being used in anticipated ways. The contribution of milk is a personal investment that differs from monetary forms of donation in terms of its intimacy. Any suspicion that part of oneself might be misappropriated and misapplied fractures the trust on which any contribution network must depend.

Concerns about the commercial applications of donor human milk usage and research also need to be addressed. If commercial gain is perceived to be a significant motivator in the world of donor milk, rather than either distribution for clinical (and nutritional) exigency and/or research leading to clear clinical (and nutritional) benefits, then a degree of cynicism threatens to corrode the basis of trust needed for effective campaigning. However, with a degree of transparency and openness about the complimentary roles of clinical distribution and research, a significant positive development can be noted. If it is understood that all milk deemed unsuitable for clinical use has a research value, then a positive donor recruitment campaign can organize itself around the slogan "not a drop wasted". The perception of wasted milk can be psychologically very troubling and the perception that one's milk is "not good enough" can also be very wounding. Pressure on individual would-be donors to somehow "make the grade" can be alleviated within a continuum of clinical and research usage which ensures that indirect as well as direct benefits to infants in need can be demonstrated.

Issues concerning the perception of milk "quality" connect with related issues to do with various technological innovations, including MyMilkLab, which offers the service to mothers of conducting detailed microanalyses of their MOM for them to be able to "understand" what is in their milk. A mother contact MyMilkLab online, a kit is sent to the mother, who then sends it back to MyMilkLab by post. Samples of MOM are then sent for analysis to a location in Israel. Clearly, this is linked to mothers wanting to trust in their own milk. A mother not having confidence in her own milk is part of the reason some mothers stop feeding their infants. In addition, it is argued that no special diet is needed for breastfeeding mothers, but this assumes a reasonably healthy diet and does not accommodate circumstances when a mother ingests harmful substances. Women who choose to donate to a donor human milk bank are screened for health and lifestyle issues. For instance, since nicotine passes through breastmilk, women who smoke are not allowed to donate to most donor human milk services. Alcohol consumption is not necessarily prohibited, although time between consumption and donation is often stated, and most DHM banks use a closed pasteurization system so that any alcohol present in the DHM will not be released during this process. Initiatives such as MyMilkLab.com, while professing to offer "reassurance" to donor mothers, may result in putting additional pressure on individuals—whereas a "no drop wasted" strategy of integrated clinical and research transparency suggests a far more productive and stress-free environment for donor milk recruitment.

The issue of whether human milk should or should not be pasteurized is one that can and should take place within an ongoing research environment. The process of pasteurization of donor human milk kills off components that may have particular research value. For example, at present, failed bacteriology following pasteurization is thrown away rather than used for research. In a global world where breastmilk is touted as "liquid gold" (Carroll 2014), while at the same time milk from other mothers is considered to have "yuck factor" (Shaw 2004), the loss of this milk is often seen as a terrible shame but part of the need to be able to guarantee secure trust in the milk.

References

Andreas, Nicholas J., Beate Kampmann, and Kirsty Mehring Le-Doare. 2015. "Human Breast Milk: A Review on Its Composition and Bioactivity." *Early Human Development* 91 (11): 629–35.

Atkins, Peter J. 2000. "The Pasteurisation of England: The Science, Culture and Health Implications of Milk Processing, 1900–1950." In *Food, Science, Policy and Regulation in the 20th Century*, edited by D. Smith and J. Phillips, 37–51. Routledge.

———. 2010. *Liquid Materialities: A History of Milk, Science and the Law*. Farnham: Ashgate.

———. 2016. *Liquid Materialities: A History of Milk, Science and the Law*. London: Routledge. epublication.

Ballard, Olivia, and Ardythe L. Morrow. 2013. "Human Milk Composition: Nutrients and Bioactive Factors." *Pediatric Clinics of North America* 60 (1): 49–74. PMC. Web. July 6, 2018.

Blais, D. R., J. Harrold, and I. Altosaar. 2006. "Killing the Messenger in the Nick of Time: Persistence of Breastmilk sCD14 in the Neonatal Gastrointestinal Tract." *Pediatric Research*, 59: 371–76.

Bode, L. 2015. "The Functional Biology of Human Milk Oligosaccharides." *Early Human Development* 91 (11): 619–22.

Bourlieu, C., and M. C. Michalski. 2015. "Structure-Function Relationship of the Milk Fat Globule." *Current Opinion in Clinical Nutrition & Metabolic Care* 18 (2): 118–27.

Boyer, Kate and Maia Boswell-Penc. 2010. "Breast Pumps: A Feminist Technology, or (yet) 'More Work for Mother'?" In *Feminist Technology*, edited by Linda L. Layne, Sharra Louise Vostral, and Kate Boyer. Urbana: University of Illinois Press.

Briere, C. E., J. M. McGrath, T. Jensen, A. Matson, and C. Finck. 2016. "Breast Milk Stem Cells: Current Science and Implications for Preterm Infants." *Advances in Neonatal Care* 16 (6): 410–19.

Cannon, A., F. Kakulas, A. Hepworth, C. Lai, P. Hartmann, and D. Geddes. 2015. "The Effects of Leptin on Breastfeeding Behaviour." *International Journal of Environmental Research and Public Health* 12: 12340–55.

Capuco, A. V., and R. M. Akers. 2009. "The Origin and Evolution of Lactation." *Journal of Biology* 8 (4): 37.

Chen, X. 2015. "Human Milk Oligosaccharides (HMOS): Structure, Function, and Enzyme-Catalyzed Synthesis." *Advances in Carbohydrate Chemistry and Biochemistry* 72: 113–90.

Cheng, S. K., E. Y. Park, A. Pchar, A. C. Rooney, and G. I. Gallicano. 2016. "Current Progress of Human Trials Using Stem Cell Therapy as a Treatment for Diabetes Mellitus." *American Journal of Stem Cells* 5 (3): 74–86.

Choi, S. S., S. R. Lee, and H. J. Lee. 2016. "Neurorestorative Role of Stem Cells in Alzheimer's Disease: Astrocyte Involvement." *Current Alzheimer Research* 13 (4): 419–27.

Clarke, Joseph. 1790. "Observations on the Properties Commonly Attributed by Medical Writers to Human Milk, on the Changes It Undergoes in Digestion, and the Diseases Supposed to Originate from this Source in Infancy." *The London Medical Journal* 11 (Pt. 1): 71–91.

Cossey, V., Vanhole, C., Eerdekens, A., Rayyan, M., Fieuws, S., and Schuermans, A. 2013. "Pasteurization of Mother's Own Milk for Preterm Infants Does Not Reduce the Incidence of Late-Onset Sepsis." *Neonatology* 103 (3): 170–76. www.karger.com/DOI/10.1159/000345419

Cregan, M. D., Fan, Y., Appelbee, A., et al. 2007. "Identification of nestin-positive putative mammary stem cells in human breastmilk." *Cell Tissue Research* 329: 129–36.

Daniels, B., A. Coutsoudis, C. Autran et al. 2017. "The Effect of Simulated Flash Heating Pasteurisation and Holder Pasteurisation on Human Milk Oligosaccharides." *Paediatrics and International Child Health* 37: 204–209.

Daniels, B., S. Schmidt, T. King, K. Israel-Ballard, K. Amundson Mansen, and A. Coutsoudis. 2017b. "The Effect of Simulated Flash-Heat Pasteurization on Immune Components of Human Milk." *Nutrients* 9 (2): 178. http://doi.org/10.3390/nu9020178

Dasgupta, Shreya. 2015. "Earth. Origins. Milk. Why did some animals evolve milk and breastfeeding?" http://www.bbc.com/earth/story/20150725-breastfeeding-has-ancient-origins.

Donovan Bernard, T. 2018. "Lactation." *Encyclopædia Britannica*. Encyclopædia Britannica. Accessed July 6, 2018. www.britannica.com/science/lactation.

Dror Daphna, K. and Lindsay H. Allen. 2018. "Overview of Nutrients in Human Milk." *Advances in Nutrition* 9 (suppl_1), May 1: 278S–294S. https://doi.org/10.1093/advances/nmy022

Escherich, Theodor. 1884. "Klinisch-therapeutische beobachtungen aus der choleraepidemie in Neapel." Mun Med Wochenschrift 31: 561–64.

———. 1885a. "Ueber die bacterien des milchkothes, Artz Intelligenz-Blatt." *Mün Med Wcsht* 32: 243, 12.

———. 1885b. "Die darmbakterien des neugeborenen und säuglings." Fortsch der Med 3: 515–22, 547–54.

———. 1886. "Die darmbakterien des säuglings und ihre beziehungen zur physiologie der Verdauung." Stuttgart Ferndinand Enke 13.

———. 1988 [1884]. "The Intestinal Bacteria of the Neonate and Breast-Fed Infant." *Reviews of Infectious Diseases* 10: 1220–25.

———. 1989 [1885]. "The Intestinal Bacteria of the Neonate and Breast-Fed Infant." *Reviews of Infectious Diseases* 11: 352–56.

Escuder-Vieco, Diana, Irene Espinosa-Martos, Juan M. Rodríguez, Nieves Corzo, Antonia Montilla, Pablo Siegfried, Carmen R. Pallás-Alonso, and Leónides Fernández. 2018. "High-Temperature Short-Time Pasteurization System for Donor Milk in a Human Milk Bank Setting." *Frontiers in Microbiology* 9: 926. www.frontiersin.org/article/10.3389/fmicb.2018.00926

Fields, D. A. and E. W. Demerath. 2012. "Relationship of Insulin, Glucose, Leptin, IL-6 and TNF-α in Human Breast Milk with Infant Growth and Body Composition." *Pediatric Obesity* 7 (4): 304–12.

Goldman, A. S. 2002. "Evolution of the Mammary Gland Defense System and the Ontogeny of the Immune System." *Journal of Mammary Gland Biology and Neoplasia* 7: 277–89.

Goldman, A. S., C. Garza, B. L. Nichols, and R. M. Goldblum. 1982. "Immunological Factors in Human Milk During the First Year of Lactation." *Journal of Pediatrics* 100: 563–67.

Groer, M. and M. W. Davis. 2002. "Postpartum Stress: Current Concepts and the Possible Protective Role of Breastfeeding." *JOGN Nursing* 31: 411–17.

Grøvslien, A. H. and M. Grønn. 2009. "Donor Milk Banking and Breastfeeding in Norway." *Journal of Human Lactation* 25 (2): 206–10.

Gustafsson, L., O. Hallgren, A. K. Mossberg, et al. 2005. "HAMLET Kills Tumour Cells by Apoptosis: Structure, Cellular Mechanisms, and Therapy." *Journal of Nutrition* 135: 1299–1303.

Håkansson, A., B. Zhivotovsky, S. Orrenius, H. Sabharwal, and C. Svanborg. 1995. "Apoptosis Induced by a Human Milk Protein." *Proceedings of the National Academy of Sciences of the United States of America* 92 (17): 8064–68.

Hakansson, A. P., H. Roche-Hakansson, A. K. Mossberg, and C. Svanborg. 2011. "Apoptosis-Like Death in Bacteria Induced by HAMLET, A Human Milk Lipid-Protein Complex." *PLoS One* 6 (3): e17717.

Hallgren, O., S. Aits, P. Brest, L. Gustafsson, A. K. Mossberg, B. Wullt, and C. Svanborg. 2008. "Apoptosis and Tumor Cell Death in Response to HAMLET (Human Alpha-Lactalbumin Made Lethal to Tumor Cells)." *Advances in Experimental Medicine and Biology* 606: 217–40.

Hassiotou, F., A. Beltran, E. Chewynd, et al. 2012. "Breastmilk is a Novel Source of Stem Cells with Multilineage Differentiation Potential." *Stem Cells* 30 (10): 2164–74.

Hassiotou, F. and D. T. Geddes. 2015. "Immune Cell-Mediated Protection of the Mammary Gland and the Infant During Breastfeeding." *Advances in Nutrition* 6 (3): 267–75.

Hill, D. R. and D. S. Newburg. 2015. "Clinical Applications of Bioactive Milk Components." *Nutrition Reviews* 73 (7): 463–76.

Jiang, R., X. Du, and B. Lönnerdal. 2014. "Comparison of Bioactivities of Talactoferrin and Lactoferrins from Human and Bovine Milk." *Journal of Pediatric Gastroenterology and Nutrition* 59 (5): 642–52.

Jonas, W. and B. Woodside. 2016. "Physiological Mechanisms, Behavioral and Psychological Factors Influencing the Transfer of Milk from Mothers to Their Young." *Hormones and Behavior* 77: 167–81.

Kataev, A., O. Zherelova, and V. Grishchenko. 2016. "A Characeae Cells Plasma Membrane as a Model for Selection of Bioactive Compounds and Drugs: Interaction of HAMLET-Like Complexes with Ion Channels of Chara Corallina Cells Plasmalemma." *Journal of Membrane Biology* 249 (6): 801–11.

Latour, Bruno. 1984. *Les Microbes. Guerre et paix*, suivi de *Irréductions*, Paris, Métailié: Pandore.

———. 1988. *The Pasterization of France*. Translated by Alan Sheridan and John Law. Cambridge, MA: Harvard Publishers.

Lepore, Jill. 2009. "Baby Food. If breast is best, why are women bottling their milk?" New Yorker. https://www.newyorker.com/magazine/2009/01/19/baby-food

Lilly, M. A., M. F. Davis, J. E. Fabie, E. B. Terhune and G. I. Gallicano. 2016. "Current Stem Cell Based Therapies in Diabetes." *American Journal of Stem Cells* 5 (3): 87–98.

Litwin, S. D., B. D. Zehr, and R. A. Insel. 1990. "Selective Concentration of IgD Class-Specific Antibodies in Human Milk." *Clinical and Experimental Immunology* 80: 262–67.

Mayerhofer, E. u. Přibram, E. 1909a. "Ernährungsversuche mit konservierter Frauenmilch. Wiener klin. Wochenschr." Nr. 26. (Feeding Experiments with Conserved Woman's Milk).

Mayerhofer, E. u. Přibram, E. 1909b. "Über Ernährung mit konservierter Frauenmilch. Verhandl. d. 26. Vers. d. Gesellseh. f. Kinderheilk." in Salzburg (1909) 99. (On Food Including Canned (or Conserved) Woman's Milk).

Miralles, O., J. Sanchez, A. Palou, and C. Pico. 2006. "A Physiological Role of Breast Milk Leptin in Body Weight Control in Developing Infants." *Obesity* 14: 1371–77.

Moro, Guido E., Claude Billeaud, Buffin Rachel, Javier Calvo, Laura Cavallarin, Lukas Christen, Diana Escuder-Vieco, et al. 2019. "Processing of Donor Human Milk: Update and Recommendations From the European Milk Bank Association (EMBA)." *Frontiers in Pediatrics*. https://www.frontiersin.org/article/10.3389/fped.2019.00049.

Mosca, F. and M. L. Giannì. 2017. "Human Milk: Composition and Health Benefits." *La Pediatria Medica E Chirurgica* 39 (2). https://doi.org/10.4081/pmc.2017.155

Oftedal, O. T. 2002a. "The Mammary Gland and Its Origin During Synapsid Evolution." *Journal of Mammary Gland Biol Neoplasia* 7 (3): 225–52.

———. 2002b. "The Origin of Lactation as a Water Source for Parchment-Shelled Eggs." *Journal of Mammary Gland Biol Neoplasia* 7 (3): 253–66.

Peila, C., G. E. Moro, E. Bertino, L. Cavallarin, M. Giribaldi, F. Giuliani, . . ., A. Coscia. 2016. "The Effect of Holder Pasteurization on Nutrients and Biologically-Active Components in Donor Human Milk: A Review." *Nutrients* 8 (8): 477. http://doi.org/10.3390/nu8080477

Picaud, J. C., and R. Buffin. 2017. Human Milk-Treatment and Quality of Banked Human Milk. *Clinics in Perinatology* 44 (1): 95–119.

Pickering, L. K., and S. Kohl. 1986. "Human Milk Humoral Immunity and Infant Defense Mechanisms." In *Human Milk in Infant Nutrition and Health*, edited by R. R. Howell, F. H. Morriss, and L. K. Pickering, 123–40. Springfield, IL: Thomas.

Rogier, E. W., A. L. Frantz, M. E. Bruno, L. Wedlund, D. A. Cohen, A. J. Stromberg, C. S. Kaetzel. 2014. "Lessons from Mother: Long-Term Impact of Antibodies in Breast Milk on the Gut Microbiota and Intestinal Immune System of Breastfed Offspring." *Gut Microbes* 5 (5): 663–68.

Savino, F., A. Sardo, L. Rossi, S. Benetti, A. Savino, and L. Silvestro. 2016. "Mother and Infant Body Mass Index, Breast Milk Leptin and Their Serum Leptin Values." *Nutrients* 8: 383.

Shaw, Rhonda. 2004. "The Virtues of Cross-Nursing and the Yuk Factor." *Australian Feminist Studies: Special Issue on Cultures of Breastfeeding* 19 (45): 287–99.

Shenker, Natalie. 2017. "The Mysteries of Milk." *The Biologist* 64 (3): 10–13.

Shulman, Stanford T., Herbert C. Friedmann, and Ronald H. Sims. 2007. "Theodor Escherich: The First Pediatric Infectious Diseases Physician?" *Clinical Infectious Diseases* 45 (8): 1025–29. https://doi.org/10.1086/521946

Strathearn, L. 2011. "Maternal Neglect: Oxytocin, Dopamine and the Neurobiology of Attachment." *Journal of Neuroendocrinology* 23 (1): 1054–65.

Thorley, Virginia. 2011. "The dilemma of breastmilk feeding." *Breastfeeding Review: Professional Publication of the Australian Breastfeeding Association.* 19 (1): 5–7.

———. 2010. "The Dilemma of Breastmilk Feeding." *Breastfeeding Review* 19 (1): 5–7.

Twigger, A. J., A. R. Hepworth, C. T. Lai, E. Chetwynd, A. M. Stuebe, P. Blancafort, P. E. Hartmann, D. T. Geddes, and F. Kakulas. 2015. "Gene Expression in Breastmilk Cells is Associated with Maternal and Infant Characteristics." *Scientific Reports* 5: 12933.

Turfkruyer, M. and Verhasselt, V. 2015. "Breast Milk and Its Impact on Maturation of the Neonatal Immune System." *Current Opinion in Infectious Diseases* 28 (3): 199–206.

Unvas-Moberg, K. 1997. "Oxytocin Linked Antistress Effects: The Relaxation and Growth Effect." *Acta physiologica Scandinavica. Supplementum* 640: 38–42.

Van Esterik, Penny. 1996. "Expressing Ourselves: Breast Pumps." *Journal of Human Lactation* 12 (4): 273–74.

Vargas-Martínez, F., R. J. Schanler, S. A. Abrams, K. M. Hawthorne, S. Landers, J. Guzman-Bárcenas, O. Muñoz, T. Henriksen, M. Petersson, K. Uvnäs-Moberg and I. Jiménez-Estrada. 2017. "Oxytocin, a Main Breastfeeding Hormone, Prevents Hypertension Acquired in Utero: A Therapeutics Preview." *Biochimica et Biophysica Acta* 1861 (1 Pt A): 3071–84.

Vogel, H. J. 2012. "Lactoferrin, a Bird's Eye View." *Biochemistry and Cell Biology* 90 (3): 233–44.

Wilson, E., K. Christensson, L. Brandt, M. Altman, and A.-K. Bonamy. 2015. "Early Provision of Mother's Own Milk and Other Predictors of Successful Breast Milk Feeding after Very Preterm Birth: A Regional Observational Study." *Journal of Human Lactation*, 31 (3): 393–400.

Winberg, J. 2005. "Mother and Newborn Baby: Mutual Regulation of Physiology and Behaviour – a Selective Review." *Developmental Psychobiology* 47 (3): 217–29.

4 "It's Not Rocket Science"
Practice and policy in human milk banking

The first bank I visited during our period of research was a hospital-based service, both of which are in major maternity hospitals in different parts of the UK, and in both cases, the first encounter was with a coffee/snack area, where visitors can purchase a tea/coffee, etc. On my first day in the field, shortly after 8am, I decided to stop in the coffee/tea area and observe the entrance to the hospital for a few minutes. A man clearly dressed in a blood bikers uniform carrying a cooler box walked past me, and I stopped him and asked, "Please excuse me, but could I ask you if that is donor human milk you have in the cool box?"[1] He answered affirmatively and asked if I was going to donate, and I told him about our study, which he said was wonderful. He told me[2] he was 73 years old and that he would keep doing voluntary bike deliveries of human milk until they make him stop driving. He also then told me that a few of the consultants in the hospital were not in favour of the use of donor human milk and asked me if I could try to have my research help them to change their minds, because he felt it was a "great" service and that it was "very important". After he left, I asked at the front desk if they could call up to the donor human milk bank to tell them I was there, and then I proceeded to go to the donor human milk bank area itself.

"It's not rocket science" is a phrase I heard used several times from the first visit onwards by one of the managers of one of the community-based sites involved in our study. Clearly, she was trying to indicate that the numerous daily tasks involved in running a large and successful donor human milk bank service were not necessarily complicated. But I could not help but feel that although each of these individual tasks in and of themselves may not be complicated, the end results in many ways could be, quite literally for some, a matter of life and death. Synchronizing such a variety of tasks did indeed appear to be no mean feat. Eventually, over the year I spent visiting each of these large donor human milk bank services, I was to experience all aspects of collecting, screening, processing, storing and distributing donor human milk in each of these banks each and every day. And although these practices are guided by the NICE guidelines, as we discussed in the previous chapter, like any organization, each donor human milk bank service has its own culture and ultimately does things its own way, which is linked to the service needs which are unique for each one. In addition, donor human milk bank services are dynamic and therefore continually changing, and although the core services generally stay the same, many things may and do change, some of which we captured in our yearlong observational study of these key large donor human milk services.

However, it is important to remember that our discussion should be considered a model based on these observations and that due to the dynamic nature of the world of donor human milk services across the UK, there is no such thing as any one representative milk bank. One of the milk banks involved in our study produced a YouTube video which begins with the caption "Everything you need to know about the Human Milk Bank".[3]

A major part of the overall research involved becoming familiar with the everyday processes and procedures of donor human milk banking services across the UK—research which is particularly relevant because one of the research questions identified in the NICE guidelines is "What is the effect of the process of milk banking on the nutritional and immunological components of donor milk?"[4] Understanding process and practice in the everyday life of the four donor human milk services involved in our study were a key part of the 12 months of observational work conducted as part of this study. Using classical field observations, including extensive field notes (which were anonymized) form the data used for this chapter.

Ethnographic methods were originally linked to the social sciences, and in particular, the Chicago School of Sociology[5] (Atkinson et al. 2001) also had links to early Chicago women (in particular Annie Marion Maclean) and to research related to immigration, gender and work—all topics which helped to frame our discussion in this chapter (Hallett and Jeffers 2008; Deegan 2014). Our ethnographic data provides a detailed description of the "process of milk banking" in the four milk banks involved in our study, and these descriptions form important parts of understanding the worlds of donor human milk services. In this chapter, we explore the everyday life of a donor human milk bank, particularly from the perspective of the staff involved in this health service. Using our field note data to frame our discussion, we offer a hybrid journey through a day in the life of a donor human milk service. By considering "time" and "space", two highly theorized concepts within the social sciences and the humanities, we offer interpretations of the ways in which time and space shrink within the worlds of human milk banking. Organized around the definition of a milk bank, we discuss processing, storing, collecting, distributing and screening. The working regime, the practical environment or physical environment of the donor human milk bank itself, as well as the staffing of milk banks, create very specific social environments and interactions which normalize particular pressures and priorities.

Rooms with a view

The four donor human milk banks involved in our study, as we mentioned, were the largest in the UK at the time of our research, with two being physically located in hospitals and two located in the community. Despite the recent movement of one of these community banks to a hospital setting (a plan which was discussed throughout our fieldwork, but which did not happen until a year after our fieldwork had been completed), we use our ethnographic field note data for this chapter, and therefore we juxtaposition hospital and community-based settings. As we were told in our interviews and during our ethnographic fieldwork, there are advantages and disadvantages to either the hospital and/or the community-based settings.

The settings in all four banks involve separate rooms for administration and pasteurization, although in some cases, these are only separated by a door which when pasteurization is not being done is often kept open, whereas in another case, they are located on separate floors and the room for pasteurization is separated by controlled access and an additional preparation room. It is important to note here that human milk is not sterile, and therefore as the NICE guidelines state, clean rooms as opposed to sterile rooms are key to the best operations practice, and none of the pasteurizing rooms in our study would be considered sterile, but all are clean, with one being a dedicated pasteurizing room with no other activities occurring in that room and therefore potentially being the cleanest of the rooms.

Hospital-based milk bank services have the advantage of having the actual milk bank located on their premises, including the visibility of the service for not only staff but also for patients and their families. One of the disadvantages of the hospital-based service is the medicalized features of the human milk. The medicalization of donor human milk is a topic which has been identified in other countries (Palmquist 2014; Carroll 2014) and carries with it the potential of alienating mothers from their birthing experiences. However, we could equally argue that this medicalized need of donor milk also serves to encourage donation based on maternal empathy. No mother wishes for her own child to be hospitalized, and helping other mothers in essence helps her to feel she is doing something good for society.

One of the hospital donor human milk services involved in our study is located on the same floor as the NICU but on the opposite side near the parents' rooms. The other hospital setting is linked to the infant feeding advisor area and is intimately tied to this additional service which is seen to be paramount to the supply necessary for donor human milk services. One of the community-based services also saw these links to infant feeding support as also key, with the manager having, as in the hospital setting bank, a dual role as infant feeding specialist as well. In this context, lactation consultancy and donor milk banks form a practical continuum of breastmilk provision. This community-based service is located near a community health centre, although it has now moved to a hospital. The other community-based bank actively decided that this was not part of the service and was better served by specialists in infant feeding support from communities or hospital settings. This second community service is no longer physically located near a hospital, although they have depots connected to hospitals, which allows for potential 24-hour delivery of milk.

As we have previously mentioned, all four of the donor human milk services involved in our study are part of the UK NHS, meaning that the services are governed within the NHS and therefore have structures in place to ensure safety and quality. Significantly, the NHS celebrated its 70th anniversary in 1948, having started before the end of the Second World War, and is now stated to be "one of the largest employers in the world" with "over 1.5 million staff" which include both clinical and non-clinical staff, but does not count the numerous volunteers on which the system also relies,[6] we will now briefly discuss the key staff involved in all of the donor human milk services involved in our research.

Making a difference

The MUIMME project was designed in close collaboration with the four milk bank managers (although one prefers the term co-ordinator) who acted as gatekeepers to the other aspects of the service. One manager retired before ethnographic data collection could begin, but she continued to be very supportive and to help with the rest of the data collection. In three of the banks, the day-to-day workings of the bank were performed by three members of the staff, although in two services, one of these members of the staff was only part time, and in two others, the manager was also seconded to infant feeding, an issue that not all services felt was part of the necessary job descriptions for donor human milk bank staff. Once the equipment has been covered, the largest cost associated with the processing of donor human milk is staffing costs, so it is clear that there are a lot of efforts towards wishing to reduce these costs as much as possible. In addition to the core members of the staff running the service on a daily basis, additional support is given through microbiological staff. There are also a number of volunteers affiliated with each of the service, as we will discuss more in a moment. But it is also important to remember that these volunteers are not cost free, since paid staff are needed to perform the necessary administrative duties required to coordinate these volunteers.

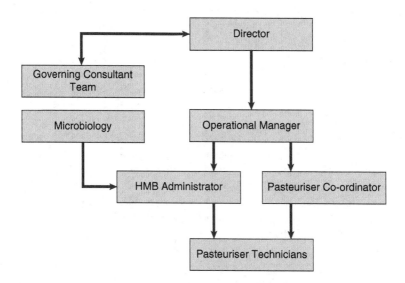

Figure 4.1 Administrative Structure
Source: Fieldnotes from the Northwest Human Milk Bank, Chester, UK.

With so few staff involved in these services, it is no wonder that the abilities to perform the core tasks are understood by all members of the staff. Accordingly, the administrative duties, which include not only interacting with donors and potential donors but also being able to take a history, are known by all staff. Administrative duties also include record keeping and more general public relations. Some banks argue that in addition to manager/co-ordinator, donor human milk bank services need a director who oversees the planning, implementation and evaluation of the service as a whole, which is the model that one of the services involved in our study adopted during our research. An ideal organizational structure for staff could involve seven people, including additional microbiological staff (Figure 4.1).

In addition to a year of our observational work in each of the banks, we conducted interviews with up to eight members of the staff from each milk bank service.

Paths to prime processing

Hollie[7] told us her story for our research, and she also told her story to a local newspaper (Williams 2018). Part of the pasteurizing staff at one of the community banks involved in our research, she is also the mother of a prematurely born infant who received donor human milk, which is directly connected to the reason why she works at the milk bank. Like others involved in donor human milk banking services across the UK, she has a very personal investment in this service, seeing it as an important part of her firstborn child's early healthcare. Being hired as a dedicated pasteurizing or processing member of staff is one of the key everyday activities associated with donor human milk services.

It is before 6am, but generally hospitals never close, and in some ways, neither do the donor human milk services, in particular those donor human milk services located in hospitals, or which have depots in hospitals, but in all cases, part of the organization of the day-to-day running of a donor human milk service is about being able to make sure that milk is always available to those who are in need. For most of the services, the morning is often devoted to the pasteurization process, although in two services with staff dedicated to pasteurization, pasteurization occurs throughout the day. The pasteurizing process, as we mentioned in the last chapter, is a feature of donor human milk in the UK, although in some rare cases non-pasteurized milk is used, such as in Norway (Grøvslien and Grønn 2009) or some parts of Germany (Springer 1997).

Each of the banks also have staff members whose main job description is pasteurization of the milk and who are sometimes called technicians. In some cases, this involves one person being in a more managerial position of responsibility, while at the same time other members of the pasteurizing staff may hold less responsibility. But a key part of this role involves surveillance of microbiological results, although often the manager/co-ordinator is the key individual linked to this responsibility, and in some cases, more than one person is needed to make this final call. In the pasteurizing room, there is also often a need to sterilize and maintain hygiene, which takes substantial time and is often integrated into the processing of the milk. In all of the banks involved in our study, there are dedicated staff involved in the milk processing, although, as we mentioned, all staff members are

78 *"It's Not Rocket Science"*

able to perform the pasteurization process if necessary. Several of the milk banks produced online discussions of how the milk is processed, and these have been used to organize our discussion.[8,9]

Step 1 Defrosting

The evening before the milk is to be pasteurized, it is put in a dedicated refrigerator to defrost overnight. This part of the processing involves not only the pasteurizing staff but also the administrative staff, since milk is prepared for processing based on organizing the milk that has come in and the dates it needs to be pasteurized. Staff members are always having to juggle when and how milk will begin its processing path, and this includes milk which needs to be processed immediately if it arrives at the bank, and it is beginning to be defrosted. Time and time management is a key part of this organization because the pasteurization must occur within three months after expression (Figure 4.2).

Figure 4.2 Fridge for Storing Donated Human Milk before being pasteurized
Source: Fieldnotes MUIMME project. Photo taken by Tanya Cassidy.

The earlier image of a refrigerator, or FRIDGE, as it is clearly labelled, was taken (with permission) during one of my field trips to one of the donor human milk bank services. Note that it also says, "RAW EBM", so as to clearly indicate that the milk has not been pasteurized yet.

Not all the refrigerators involved in our study are made of steel like this one, but they all have doors which are not able to allow light to pass through. Some laboratory-grade refrigerators and freezers can have glass doors, and when preparing our manuscript, it came to our attention that this could have an effect on the milk in storage by unnecessarily exposing the milk to additional light, so the preferred refrigerator and freezers do not have glass doors.

Step 2 Mixing, sieving and bottling

When the milk is fully defrosted, staff members transfer the milk to a "jug", which, as we see in the image that follows, is steel, but may be hospital-grade plastic. In the UK, unlike in North America, NICE guidelines state that milk is not pooled between mothers but can and is pooled for a single mother. Logistically, this means that more than one expression from a particular donor is pooled into one processing session as this increases and distributes the potential fat and other contents of a donor's milk throughout the various donations since we know that milk is extremely dynamic and can be dependent on a number of factors.

In some banks, the milk in the jug is then given a light stir. Not all of the banks mix the milk, as some feel it is better to not disturb the milk as much as possible. All the banks involved in our research, as we will discuss more in the next chapter, send single-use sterile plastic containers to the donors to collect their milk in. However, sometimes expressions which have been done prior to becoming a donor are also accepted, but some banks do not accept donations which have been collected in breastmilk bags. I was told that this is because the bags keep too much of the fat content of the milk and can be more difficult for staff to handle. But some of the banks still feel it is important to accept milk which has been expressed in bags, although they want donors to use the sterile containers they send for ongoing donations.

Both food-industry and medical-grade materials are used by staff members to process these materials. Although staff stress that human milk is not sterile, to limit additional contamination, this part of the process is conducted on a clean airflow table, similar to the ones used in biology labs, and is produced by the same company that makes the pasteurizing machine (Figure 4.3).

Step 3 Microbiology

A sample from each "batch" is sent for microbiological screening, but the day-to-day staff must prepare the sample to be sent. The results are communicated again to the staff, often as part of administrative duties. If the sample is contaminated, the batch is discarded by the staff, and this was something which occurred at each of the banks during our data collection, although the failure rates vary from bank to bank. These failure rates are monitored, and if deemed necessary, checks and

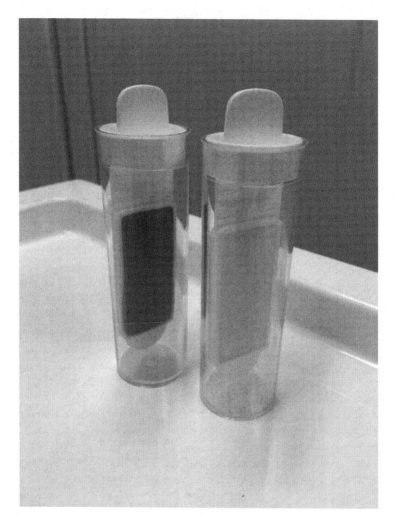

Figure 4.3 Bacteriological Dipslides
Source: Provided with kind permission by Debbie Barnett at the One Milk Bank for Scotland.

policies are implemented, including contacting donors about best practices for hygienic donations—a highly sensitive and difficult communication and something we will discuss more in the next chapter.

Sending the samples to the labs is one of the areas where the services in our study differ the most. In most cases these tests are performed by the NHS laboratory staff affiliated with the service. In the case of the two community-based banks, these samples are sent by internal post, and any delays could lead to potential negative tests, resulting in milk being unnecessarily discarded. Delays in transport was also a concern for one of our hospital-based services, and as a result, they took steps to use a dipstick which is not affected by delays in transport.

The team ran comparisons tests with traditional procedures and found that the results were essentially the same, but unfortunately did not publish these results. In addition, since these tests use less milk, a policy decision was made to use dip slides for microbiological tests. Although these slides cost more than the initial costs for traditional laboratory tests, when factoring in staff time involved in traditional testing procedures which are significantly saved with the dip slides, the cost is negligible and was determined to be far better. Incidentally, it was considered inappropriate to dip the slide in the full jug of milk, so small amounts are removed and put on the slide, and then the slides are sent to labs in the hospital that determine the results, which are returned to the donor human milk bank staff as negative or positive.

Step 4 Analysing milk for macronutrients

In two of the banks involved in our research, each batch is tested for its macronutrient value (fat, protein, carbohydrate, total solids and calories). As we mentioned, all of the milk banks involved in our research are research active, and as part of this, three of them have actively used milk analysers, in particular the main one available on the market, and have even participated in marketing of this product. However, one bank does not provide this information to clinicians for individuated milk treatment. One of the other bank's staff members told me that the staff used to provide this information, but they were told that the clinical staff did not use it. Whereas one of the other banks that actively includes it feels that it is there in case it is considered important for clinical use. Individual nutritional components are part of the advertisement for such equipment, but the operation takes up more staff time. In addition, two machines were broken during our research, and they had to be repaired, meaning that there were times when this information was not available, so if clinical decisions were dependent on this information, such a delay would be a problem.

Step 5 Pasteurizing

Before going into the pasteurizing machine (Figure 4.4), the bottles of milk are placed into metal baskets or grates, and put into the pasteurizer. Most of the pasteurizers have the ability to have double baskets and to hold bottles of different sizes (50 mls, 100 mls, 200 mls), so administrative decisions regarding the size of bottles needed for each batch are also part of the staff duties.

The two main pasteurizing systems used in donor human milk bank services across the UK use the Holder pasteurization method, which we discussed in the last chapter, but one involves immersion of the bottles in water, and therefore the bottles have a special heat seal. This means there is an additional step for staff members prior to putting the bottles in the baskets for pasteurization. The immersion method, however, cools the milk down to 4 degrees, allowing it to be immediately put in the freezer after it is taken from the pasteurizer. The other system has an additional step for staff members who are required to put the recently pasteurized milk in a fridge until it is cool enough to be then transferred to the freezer. We were told that there had been some concerns several years ago about

82 *"It's Not Rocket Science"*

Figure 4.4 Pasteurizing Machines
Source: Fieldnotes MUIMME project. Photo taken by Tanya Cassidy.

immersion, although the heat sealing seems to have alleviated many of these problems.

The two main brands of donor human milk pasteurizers look very similar on the outside. One is kept on a counter top and is physically smaller, advertising itself for smaller spaces, and is in fact used by the two physically smaller bank services involved in our study, whereas the other version is floor based and larger, and it is the main pasteurizing machine used in the other two services. Both machines involve a similar amount of staff time. Although the larger machines also completely cool the milk, which is then ready to immediate go into the freezers, whereas the smaller machines require the milk to go into a refrigerator until the milk is cool enough to go into the freezer, which again means more time spent by staff processing the milk.

Step 6 Labelling

Each chilled bottle is labelled before being put into the freezer. In two donor human milk services involved in our study (one hospital and one community based), these labels involve a unique batch and bottle number, as well as the expiry date. Additionally, one service, as we mentioned earlier, also includes the nutritional information gathered in Step 4. The other service is the only one which uses a bar coding machine and automates all of this information, which it feels reduces staff time needed to perform this important step, which is otherwise done by hand. The other two services also do the labelling by hand and use the smaller machines, thus involving additional staff time to complete this part of the processing.

During our ethnographic data collection, there was a concern that one of the companies that made the bottles for one of the pasteurizers had decided to stop production. Since the bottles for the other pasteurizer are round, they would not be able to be used in the other system, and so a concern about the availability of bottles consumed staff members for several months, resulting in one service changing pasteurizers.

Step 7 Patiently awaits

The cooled, pasteurized, labelled milk is stored in freezers until all the bacteriological tests have been cleared, and the milk has been cleared for distribution to infants in need. We will discuss the recipients more in the next chapter, but for staff members, distribution is another part of their organizational duties. Staff members are needed to determine who is in need of what milk and when. Once this is determined, special cooler bags are packed and delivered to hospitals, or on occasion to mothers in the community.

All of the banks use these boxes/bags on occasion, but some are white (Figure 4.5), green or purple. For some deliveries, one service has also started using a specialist service to deliver some of their milk, particularly if it is being delivered a long distance, but cost is always an issue regarding delivery of milk, and therefore volunteers are often employed for this part of the service.

Lunching together

Donor human milk bank services are very busy places, and the mornings often go by very quickly, and as we mentioned, they are often are dominated by the processing of the milk; meanwhile, the phones have been ringing, often before 9am. Potential donors are often new mothers whose time is dictated not by 9 to 5 hours, but instead is much more erratic and linked to on-demand feeding schedules, and as we discuss more in Chapter 5, this means that surfing the web at 3am is not an uncommon activity while the mothers are also feeding their newborns, and therefore some of the services have online materials which the potential donors can complete and email to the bank. All of these emails need to be read and responded

84 *"It's Not Rocket Science"*

Figure 4.5 White Boxes for transporting milk
Source: Fieldnotes MUIMME. Photo taken by Tanya Cassidy.

to, while at the same time the telephone also needs to be answered. Increasingly, the receptionist is also a key gatekeeper to the public, but needs to make initial assessments regarding the potential of any particular donor. Lunchtime illustrated one of the key differences between the banks. In two of the services we worked, the staff lunched together, which led to camaraderie and close ties between the staff members. One of the other services does not allow staff members to all take lunch at the same time but often has outside gatherings with the staff which also serves to create an increased sense of camaraderie.

For some of the services, the bank never stops, and so staff members take lunch separately so that there is always someone there to answer the telephones, which is part of the reason why all staff members are trained in key administrative tasks, such as dealing with potential donors, as we mentioned previously. This is especially true with regard to administrative duties, which are often performed while some of the staff members have their lunch. But this clearly seems to indicate that some of the administrative duties are the least important, although some are also the most important, particularly when dealing with donors, potential donors and public relations in general. However, several of the day-to-day administrative tasks are also performed by volunteers in some services, such as preparing packs which include sterile empty bottles, thermometers, pens, etc., to send to donors.

Invaluable volunteers

As I noted at the beginning of this chapter, on my first day of ethnographic research, the first person I met was a volunteer blood biker. In addition to the NHS paid staff members involved in all of the services we studied, three of the four banks in our research also actively use volunteers. Due to limited space, the one bank was not able to accommodate volunteers, although this service has recently moved and therefore may be able to include volunteers in the future. Volunteers perform a variety of tasks in the different banks, including picking up donations from donors, delivering milk to recipient hospitals or individuals and helping in the office by doing a variety of things, such as preparing boxes of bottles, instructions, etc., to be sent to donors. Similarly to some paid staff, volunteers often have a personal connection with wanting to help donor human milk banking or simply give back to their community in some way. However, others may be linked more generally to helping and community generosity.

We began this chapter by discussing one of the blood biker volunteers, a highly energized retired man who wanted to help milk banking and the babies it also helps. These volunteers ride their motorcycles around the UK and Ireland, helping to deliver a variety of medical products, including donor human milk. In some cases, the volunteer bikers are used to collect donations from donors, and in other cases, they are used to deliver safe pasteurized milk to hospitals or mothers in the community, but not always. As we discuss in Chapter 6, this service in fact cuts across borders, allowing collaboration between the UK and Ireland.

Boob-stock and beyond

As we mentioned earlier, all of the milk bank services involved in our research are run daily by a comparatively small number of staff members employed through the NHS, although, as we have discussed, this is heavily supplemented by volunteers. Some of these staff members are integrated into other services, especially infant feeding support, which are linked to clinical concerns about breastfeeding support.

There continues to be clinical concern that the provision of donor human milk services may impede on MOM, and therefore we see the splitting of job descriptions, with some co-ordinators also being infant feeding councillors. This integrates the importance of donor human milk services with larger social issues of encouraging MOM, in particular for vulnerable infants, but it can also be compromising in terms of the donor human milk banking staff members and their available time for working on activities associated with the service. We saw differences, as we mentioned earlier, in the integration of infant feeding support, whether for mothers of recipients or for donors themselves. Clearly, one of the advantages of a community-based service is its potential direct access to community mothers, whether they are there to support them in their breastfeeding journey or not, an aspect of the service which needs to be considered when thinking about the service. The demedicalized features of a community-based human milk service means, however, that alternative health services, such as community-based infant feeding support, are seen to be better provided by other health staff in alternative services. The links between breastfeeding and donor human milk services continues to be a complex issue that has cultural links, as we will discuss again in Chapter 6.

Notes

1 Note that consent was obtained from this volunteer blood driver, and therefore this exchange was deemed allowed to be included in our discussion.
2 Please note that consent was obtained from all participants in our study, and their contributions were anonymized.
3 See www.youtube.com/watch?v=EFIm5cu5Q-s.
4 See www.nice.org.uk/guidance/cg93/chapter/2-Research-recommendations.
5 The word "ethnography" is derived from the Greek ἔθνος (*ethnos*), meaning "a company, later a people, nation" and -graphy, meaning "writing" (see www.etymonline.com/word/ethno-). According to the *Oxford English Dictionary*, ethnography can be traced back to at least the eighteenth century and to travel narratives. Today, it is widely recognized that there are many forms of ethnography, from critical ethnography, realist ethnography, to autoethnography, all of which inform our discussion.
6 See www.nhs70.nhs.uk/get-involved/support-the-nhs/join-the-nhs-team/.
7 Please note that this is her real name since she chose to discuss this information in an openly available online article (please see Williams 2018).
8 See www.northwesthmb.org.uk/about/how-milk-is-processed/.
9 See www.youtube.com/watch?v=wH2jw5vXj2s.

References

Atkinson, P., A. Coffey, S. Delamont, J. Lofland and L. Lofland. 2001. *Handbook of Ethnography*. New York: SAGE Publications.
Carroll, Katherine. 2014. "Body Dirt or Liquid Gold? How the 'Safety' of Donated Breastmilk is Constructed for Use in Neonatal Intensive Care." *Social Studies of Science* 44 (3): 466–85.
Deegan, Mary Jo. 2014. *Annie Marion MacLean and the Chicago Schools of Sociology, 1894–1934*. New Brunswick: Transaction Publishers.
Grøvslien, A. H. and M. Grønn. 2009. "Donor Milk Banking and Breastfeeding in Norway." *Journal of Human Lactation* 25 (2): 206–10. https://doi.org/10.1177/0890334409333425
Hallett, Tim and Greg Jeffers. 2008. "A Long-Neglected Mother of Contemporary Ethnography: Annie Marion MacLean and the Memory of a Method." *Journal of Contemporary Ethnography* 37 (1): 3–37.
Palmquist, Aunchalee. 2014. "Demedicalizing Breastmilk: The Discourses, Practices, and Identities of Informal Milk Sharing," In *Ethnographies of Breastfeeding: Cultural Contexts and Confrontations*, edited by Tanya Cassidy and Abdullahi El-Tom. London: Bloomsbury Academic.

Springer, S. 1997. "Human Milk Banking in Germany." *Journal of Human Lactation* 13 (1): 65–68.

Williams, Kelly. 2018. "Mum of Premature Baby So 'Overwhelmed' by Human Milk Bank Donations that Helped Daughter – She Got a Job There." www.dailypo Stco.uk/news/north-wales-news/mum-baby-human-milk-bank-14217383

5 Pumping for preemies[1]

Nadia[2] was born at 34 weeks and 5 days weighing 2 lb 10 oz. (approximately 1,191 grams). I had severe pre-eclampsia, so she was delivered by caesarean section. I was quite poorly and on a lot of medication, and because of that, she wasn't able to have my milk at first. When I came round from my operation, I was approached and asked if I would be willing to let her have donor milk, and at that point, I'd never even considered or heard about the human milk bank. But the hospital gave me loads of information about the process and how the milk was screened, and how it would be better for Nadia if she had breast milk rather than formula, so my husband and I decided she should have it.

So many of the mothers of recipients with whom we spoke with in our study wanted to give something back to the world of donor human milk banking like Nadia's mother earlier; she was not able to become a donor, despite considering it and having a reasonable supply of milk (Williams 2018). As we discussed in the previous chapter, she, like others in our study, eventually became employed at one of the banks where our research was conducted, illustrating something of the nature of exchange, showing how we feel a sense of obligation and therefore want to, in some way, give back, and how this is linked to important features of reciprocity, which Mauss (1925, 1954, 2002, 2016) noted forms the basis of society.

A major underlying argument within the world of donor human milk banking services is that donor human milk should never be considered a replacement for formula (Meier et al. 2017), but instead should always be regarded as a support system to help mothers to establish their own lactation. This is a point that is being emphatically restated in the context of the global expansion of donor human milk services (Cassidy and Dykes forthcoming; DeMarchis et al. 2017; PATH 2013). As we discuss in the next chapter, donor human milk services have been shown to lead to increased rates of exclusive breastfeeding (Adhisivam et al. 2017). Problems occur when donor human milk is seen as an alternative to formula, and informed parties need to actively challenge any such perception, since, as we discussed in Chapter 3, human milk is a live substance, intimately linked not only to both the mother and the infant for whom it is originally biologically linked but also to other mothers and other infants. There is evidence that cup feeding, or spoon feeding, may be preferable to a bottle, even for preterm infants (World

Health Organization 2018, 28). It is not uncommon for preterm infants to learn to breath, suck and swallow using a bottle. Therefore, we may need to change the mindset that bottle feeding should not always be assumed to mean formula feeding but can instead provide the means through which some parents can feed their infants expressed human milk, ideally their own, but if necessary from other mothers. The world of breastfeeding needs to increasingly recognize the complexity and diversity associated with feeding infants mothers' milk, either from their own or other mothers. Mothers who for a variety of reasons exclusively pump to feed their infants may feel alienated by some breastfeeding support groups (O'Sullivan et al. 2018; Cassidy 2016).

In this chapter, we explore how donor human milk banking involves establishing a maternal community of generosity and reciprocity, which is an integral part of the sense of obligations associated with this extremely personal form of exchange. Accordingly, rather than have a separate chapter on donors and one on mothers of recipients, we are integrating the two. This chapter argues that mothers of recipient babies are not the passive beneficiaries of milk banking but have contracted into a network of relationships governed by a common sense of an urgent need for human milk for human babies. The motives, experiences and discoveries of donor and recipient mothers form the heart of this chapter.

Narrative interviewing and the power of stories

The world of donor human milk services is full of "milk stories", primarily from mothers, and, as we have discussed, the matricentric feminist theoretical explorations (O'Reilly 2016) currently taking place within the world of mothering and motherhood studies are particularly helpful and interesting for our discussion. As part of our research, we planned to talk with 15 donors from each of the four milk bank services involved in our study, but in each milk bank, we had several more women volunteer to tell us their stories about donor human milk banking, with the consequence that we not only formally conducted 60 narrative interviews (Jovchelovitch and Bauer 2000) with donors but we also collected narratives from an additional 30 donors. One father participated in one of the interviews with a donor, and he told us his story of donor human milk banking as well. We had not integrated this paternal story into our original design, but plan to do so in the future. We had also planned to conduct narrative interviews with 15 parents of recipients, and for comparative purposes, we primarily asked for mothers from each of the milk bank services as well, but these proved more difficult to obtain. The Scottish and the Irish services keep searchable records that could be used to contact participants who are parents of recipients. Parents of recipients were obtained from the London and Chester services through networking connections from the manager and staff from affiliated hospitals. Ultimately, we obtained narrative interviews with 30 parents of recipients from across the UK. Not integrating parents of recipients into the world of donor human milk banking is a policy recommendation we made to advocacy groups, such as UKAMB.

In the two hospital-based services involved in our study, although both researchers observed parents of recipients on the units and talked to them informally, we made the ethical decision to not actively recruit these parents. It should be noted that none of these parents asked to be part of the study, although we were happy to tell them about the research if they wished. We know that although other researchers made a different choice (Carroll 2014a, 2014b), we felt recruiting to be part of a research project was too much of an imposition for parents undergoing such high levels of stress associated with having a hospitalized infant. As a result, we only obtained seven interviews with parents of recipients, but we also interviewed representatives of a group we had not anticipated, specifically mothers who began as parents of recipients and who either were then able to increase their own supply so much that they became donors or who following subsequent births had increased supplies and then became donors. We conducted seven interviews from this group from each of the donor milk services. Each interview lasted approximately one hour. It is important to remember that we were not attempting to conduct any kind of representative sample with these interviews but were instead conducting qualitative interviews to give us details and a depth of experience which might not otherwise be available to researchers.

As part of our research, we conducted narrative interviews with not only staff members, whom we discussed in the previous chapter, but also with donors and parents (in particular mothers) of recipients, like Nadia's mother. In each case, ethically approved information about the project was shared with potential interviewees, including the following narrative from our ethically approved *Participants Narrative Interviewing Information Guide*:

> Qualitative interviewing is purposely less structured than questionnaire-based interviews, although these may include qualitative open-ended questions. Generally, these methods are used when researchers want to obtain greater depth of information or are studying topics that have limited research. The qualitative interviewing known as "narrative interviewing" recognizes that the participants' perspective is the key feature associated with qualitative data and is designed to limit the role and influence of the interviewer as much as possible during the interview exchange. A key feature of this kind of interview concerns the interviewer being an active listener. The interviewer will listen to your "story", and ideally the interviewee will have a sense of someone extremely interested in whatever they wish to tell them about the particular topic the interviewer is interested in studying, in this case, donor human milk banking.

Please note that this narrative went on to discuss that interviews would be digitally recorded to facilitate more extensive analysis at a later date. I also explained that the interviewee was always in control and could stop the interview at any point. All interviewees were comfortable telling their stories, and no one chose to stop or change things. All interview data was anonymized, and we made choices about what to include in the following discussion. Once consent forms were signed and returned, all interviews began with, "Please tell us your story about donor human milk banking".

Moreover, many donors who had heard that we were looking for "stories about experiences with donor human milk banking" offered to speak with us, and many also wrote out their stories, or told them to others as well, and several of these stories were in fact published on one of the websites affiliated with one of the services. Many donor human milk services around the world will often publish stories from both donor and recipient perspectives, most often told by mothers. "Milk stories" are often presented on many donor milk bank web pages under a link called either "donors stories" or "recipient stories", all of which are very individual and personal, but all of which can help us to delve deeper and therefore better understand the world of donor human milk banking. Behind each of these stories are mothers and their infants, or more accurately the maternal-infant dyad, and the relationship that these two form. As we discussed in Chapter 3, this relationship is explored by immunological science, but it transcends biology, as it is about nurture in a far broader sense. As Penny Van Esterik has recently argued (Van Esterik and O'Connor 2017), these early nurturing relationships can and should be viewed as forming the foundations of societies and transmitting cultural understandings through the generations.

Milk and medicalized mothering

The earliest stories of donor human milk, as we touched upon in Chapter 2, emphasized the infants themselves, as was the case with stories associated with the origins of milk banking in the UK and the St Neots quads for whom Edith Dare collected and delivered milk. However, at that time (the 1930s), MOM was not recognized as having the importance that it does today. The best milk for medically compromised infants is their own mother's milk, which is specifically and organically designed and redesigned for them. In the UK, like in the USA, MOM in the neonatal unit is not pasteurized and is therefore a completely bioactive substance specifically designed for the baby. However, as we discussed in Chapter 3, donor human milk is pasteurized in most countries in order to help to combat any potential transmissions of infections or contaminations associated with the milk itself. This was not always the case in the UK, nor is it policy for all countries around the world. For instance, in France, in order to potentially combat cytomegalovirus (CMV) infection, milk on neonatal units is pasteurized, and therefore the MOM makes up the majority of milk in a lactorium (Picaud et al. 2018; Lopes et al. 2018).

Increasingly, we recognize the maternal-infant dyad and therefore give voice to the mothers involved in these exchanges. The mothers who shared their stories about donor human milk banking spent time talking about their babies and usually began their stories when they were pregnant and told me their birth stories as well. Such birth experiences were uneventful and "routine" for many donors, but for others, they were highly medicalized, emotional and stressful experiences. This was unfortunately, if inevitably, often the case for many of those parents whose infants were recipients of donor human milk.

> When I was 23 weeks pregnant, they discovered that I had heart failure, and so at 29 weeks, the medical team made the decision to deliver Euan because

he wasn't growing, and so they gave me steroids to help with his lungs, and then he was delivered by caesarean and weighed just under 1 kg. After they ventilated and stabilized him and then they showed him to me before they took him to the NICU. Before he was born, I knew I could have a premature baby because I had been so sick, and so I had already been told about donor milk, and I decided that if he was premature that I would use it.

Laura went on to tell me about how she had been so ill that she basically spent six days away from her newborn infant and was too sick to express for her infant at least in the earliest days. Accordingly, Euan received donor milk for the first 18 weeks of his life, and when Laura was physically able to express, she did so often. She was never able to regain her supply, but she said how grateful she was to have had the donor milk, which she felt helped Euan recover more quickly from infections he experienced during his stay in the hospital.

Another mother I spoke with is Madison who had identical twin girls who had to be born early because one was not growing well. After they were born, one received donor milk and the other was transferred to another hospital and did not receive donor milk.

Due to having to travel between two hospitals and visiting two babies, she made the decision not to express herself, especially as she said when she tried there was nothing coming, but it is clear she did not receive support to express and that the donor milk was therefore being used as an alternative to formula.

The majority of infants who receive donor human milk are hospitalized, and the most common group of recipients consists of those infants born prematurely, although some surgical infants also receive milk. In other countries around the world, donor human milk is given to infants whose mothers are unable to feed due to health-related issues, such as being HIV positive (Israel-Ballard et al. 2005, 2006). The health concerns that many of these mothers experience are often forgotten. It is vital to remember that the mothers of these recipients, like Laura, often experienced very medicalized births, which often involved surgical and/or medical interventions, all of which have been long recognized as interfering with establishing breastfeeding (Hobbs et al. 2016). Research also indicates that for mothers who give birth prematurely, there are a number of factors involved in the initiation of breastfeeding (Maastrup et al. 2014), and there is increasing evidence that an exclusively human milk diet is best for the brain development of these infants (Blesa et al. 2019).

Discovering allomaternal[3] mothering with milk

> My name is Chitra. My son, Avy, was born premature at 31 weeks gestation weighing 1.75 kg. He was born in April 2010 at Swansea. He was born in good condition. His APGAR score was good. Avy was self-ventilating in air. He was taken to the Neonatal Intensive Care Unit. His brain scans on the first day, third day and seventh day were normal.
> I started to express breastmilk a couple of hours after Avy's birth. I expressed breastmilk every three hours for Avy's feed.
> When Avy was 4 days old, he was established on full feeds on a combination of my expressed breastmilk and formula milk. After a couple of days, he was

diagnosed with severe NEC (necrotising enterocolitis), an infection of the intestines. Avy became very poorly, and he was made nil by mouth. The neonatal consultant told me to continue expressing breastmilk. I expressed breastmilk every three hours to establish my milk supply. As Avy was nil by mouth, the EBM bottles were stored in the NICU freezer.

Avy's condition worsened, and he was put on ventilator. When Avy was 8 days old, he was moved from Swansea to Cardiff. Avy continued to be nil by mouth. When he was 13 days old, 60 percent of his intestines were surgically removed. A stoma was formed with a mucous fistula to recycle the stoma losses. After a couple of months, the stoma was reversed. Avy developed short bowel syndrome.

Avy was fed my expressed breastmilk (EBM) via an NG tube. He could tolerate up to 5 ml per hour. If he developed an infection, then he would be nil by mouth. I continued to express breastmilk every three hours. The excess EBM bottles were stored in the NICU freezer. The staff at Cardiff NICU encouraged me to donate the excess milk. I contacted Chester and North Wales Human Milk Bank (now known as Northwest Human Milk Bank) and answered the milk bank's screening questions. The neonatal consultant at Cardiff sent my blood sample to the milk bank by post for screening. The first donation of milk was collected in July 2010 from Cardiff NICU and Swansea NICU.

When Avy was 5 months old, he was moved to a paediatric ward. He was there for 11 months, slowly catching up on his feeds. When he was 16 months old, he was discharged from hospital. I continued to express breastmilk every 3 hours till Avy was 2 years old.

The milk bank made the process of donating milk as simple as possible. Over a year, I donated 1,000 bottles of my expressed breastmilk to the milk bank. Each bottle contained approximately 100 ml of EBM. This milk was pasteurized and given to many babies across the UK. The story of my donation was featured in the Cheshire and North Wales Human Milk Bank 2011 calendar.

I was told about the benefits of donor breastmilk and how donating my breastmilk was one of the most precious gifts that I could offer. Donated breastmilk helps to save the lives of premature and sick babies whose mothers are unable, for many reasons, to provide them with sufficient breastmilk of their own.

It made me very happy to be able to help other babies. I didn't even know them. I didn't see them, but I knew my breastmilk saved lives.

Chitra (who asked us to use her and her son's real names) gave these details originally in her interview during our research. Our interview inspired her to want to become more involved in the expansion of donor human milk services. As a powerful patient safety advocate already, based on her own doctoral work (Acharya 2018), she announced after our interview that she wished to establish a satellite milk depot in the north-east of England, an under-serviced area. She also informed me that she had encountered resistance to the idea from healthcare providers in the area. Approximately a year after our interview, she was asked to present her story to a conference on donor human milk, and she wrote out the earlier account, which she then shared with us. She also told us that she had included a discussion of donor human milk in her doctoral research and is keen to continue to expand this work in the future, which is part of her rationale for waiving her anonymity. We expressed our willingness to respect her wishes in this respect. It is also interesting that specific milk bank services are also keen to be identified with stories which help to show their importance, an issue which has always been part of the stories of donor human milk banking services, but also of the marketing of formula, as I have already discussed in relationship to Ireland (see Cassidy 2012).

Because Chitra's son was born prematurely, her own milk is, according to some physicians, considered more appropriate (Dempsey and Miletin 2010). Some physicians would prefer that donated milk come from mothers of infants who are at similar gestational age to those infants receiving the milk, and the issue is particularly prominent among healthcare providers on the island of Ireland; hence, the reason milk services dispense colostrum, preterm and full-term milk by colour coding (see Chapter 4). In order to supply the hospital-based preterm donations, however, there is a need to obtain milk from mothers who are in hospital themselves and/or whose infant(s) are in hospital. Most often, the donors are mothers who gave birth at full term (meaning 37 weeks gestation or later). For infants born prematurely, the ability to coordinate sucking, swallowing and respiration may not be developed until 28 to 30 weeks gestation (Lau 2015). There are many discussions about the feeding journey an infant born prematurely often makes, which may begin with total parenteral nutrition (intravenous feeding, direct in a vein) to tube feeding by naso- or oro-gastric tube to finger feeding, to cup feeding and, finally, to full breastfeeding (Bergman 2010).

Mothers of preterm infants also form part of this feeding journey, which may involve expressing. Although hand expression is now recognized as key for mothers of preterm infants (Becker et al. 2016), for many preterm mothers involved in our research, expression also involved an intimate if sometimes problematic relationship with the pump itself, although some minor massage preparation may also have been discussed. As researchers have recognized, for the mother of a preterm infant, her milk is often seen as more medicine than nutrition, and she, therefore, may find herself spending long hours with a mechanical device, rather than enjoying a more natural relationship with her infant. Accordingly, for mothers who had planned to breastfeed their infants, this may lead to an additional sense of failure (Zizzo 2013). We have also found that for mothers who had never planned to breastfeed, the emotional experiences of milk in the NICU can actually lead to more positive attitudes towards breastfeeding (Cassidy 2016). For some of these mothers of preterm infants, their donations became a way to feel that they were doing more, especially when their own infants were perhaps not doing so well. This is particularly true for mothers of infants who have the worst possible outcomes and do not survive. Bereavement donation is very prevalent throughout the UK, and in Ireland particularly, as we discussed in a separate article we have written on this topic (Cassidy and Dykes forthcoming).

> I saw it in the new mum papers, and I decided I wanted to get in touch. I was lucky enough to be able to do this, and I wanted to share this. I sent an email; they did the blood test and sent the bottles for me to fill. It was as quick and easy as that.

The majority of donors we interviewed during our research, however, delivered their infants full term and donated once their supply was established.

> I had a lot of wonderful support with my first baby, and I had a lot of trouble feeding her. But with my second baby, I had a lot of supply. And this baby would not take a bottle no matter what I did, and he kept feeding at one side of the breast. I think he didn't like the inverted nipple, so I filled up a freezer and then I contacted the milk bank to donate this milk which he wouldn't take. So, they were amazing. The volunteer ambulance service came out and collected what I had. And the second time I didn't have a lot, but I phoned them up, and they said bring it, so I dropped it off at the hospital with no parking. And then I got antibiotics for something, and then I couldn't donate that milk, and then my milk supply settled down, and I did not pump for donation again. If I had been pumping anyway, I would have donated, but he would take a bottle, so I wasn't pumping for him. I would have only been pumping for donation.

As we see with Noelle's account, her donation is intimately connected with her feeding journey with her own infant. Often, donors talk about expressing while feeding their own infants, which often meant that they increased their own supplies, sometimes remarking that it was like feeding twins. There are at least two distinct kinds of donors. There are many women who donate once or twice, giving only the minimum required amount, and then there are those who choose to be continuing donors for as long as they can be donors.

Recipient/donors—the forgotten but ideal donor

> I have two children, a boy and girl. And the first time around, it went really well. And the second time around, I was worried about what if it doesn't go well. She was born full term—exactly on her due date. Initially a quite straightforward labour. We were ready to have her at nine in the morning. And then things started to go a bit wrong. It turns out she was back to front. So she got stuck. I was pushing for a few hours. And then at the last minute, they really needed to get this baby out. They took me around to theatre. They gave me a spinal, and they tried forceps. And, thankfully, I was able to deliver her with forceps delivery. But I had been pushing for about three hours, and she had quite a lot of swelling around her head. Before I was even able to hold her, they took her immediately and said they were going to have to take her and monitor her in the neonatal unit. So there was I, just emotional, after just having my baby and not having her with me. They took me back to the delivery room, and it was about five minutes later that a nurse came back from the neonatal room, and she said to me we need to feed your baby, "What had your plans been?" All along, my plans had been to breastfeed her like had fed her brother. I think I was really emotional. I'd just given birth, and she had been taken away from me, and I didn't really know what was going on. So, I said I really wanted to breastfeed her, and they said they should have given her to me right away before they took her, but they just wanted to keep an eye on her. So, it was one of the nurses that said you know we have the option of donor milk if you

don't want to give a bottle. I don't know why. Perhaps because I had breastfed Conor. That gave me such comfort. They wanted me to recover from having had the spinal. My husband went with her. For me, the option of donor milk really did content me. My husband was with her in neonatal. She was an 8-lb baby. She wasn't one of the tiny fragile babies in the neonatal. So my husband was able to feed her the donor milk. I think she had two feeds. It was about nine hours after the birth that I was allowed to go to the ward to be with her. And that was because of much persuasion on my part. They really weren't happy with me going. I said I can stand up, and I can go in the wheelchair, and they rolled me down. So, it was about nine hours later when I first held her, and I was able to feed her myself. And, thankfully, she took to it great.

I don't know why. I was so thankful. I don't have any problems with anyone bottle feeding. But it just contented me to know she was to get the donor milk. I feel a little bit guilty too. She wasn't a little tiny; she was to me a fragile baby. But she wasn't a little 1 or 2-lb baby who needed this milk. I was just grateful.

In the nine hours while I waited, there were nurses in the ward who helped me to express colostrum, and my husband was also able to give that to her.

So I decided if we are able to continue with this journey, I thought to myself, if we are able to then donate ourselves, because we are just so thankful. So, when my daughter was about 3 months old, I contacted the milk bank. I said I would be interested. I didn't know much about it. But they were super. They emailed me a lot of information. I was a bit nervous, and I was worried that I might not be able to do it. But it worked out great, and we were able to donate for just over a month to make up the minimum donation. At first, I thought I won't ever be able to make this amount, but we were. The reality is it was increasing my supply, and my daughter was getting more than she needed.

Joan then went on tell us about her sister who had a 6-month-old baby at the time of our interview. She had breastfed two other infants prior to having her third, and with this third baby, she had also donated to the milk bank. Several donors talked about how they would tell friends and family members, and that often this would result in other people becoming donors.

Death and donation

My neighbour's niece was pregnant and got meningitis and died. Because she was quite late on in her pregnancy—I think it was 28, 30 weeks—the baby was able to be saved. Obviously, it was a really sad situation. and the baby had donor milk. I was really moved by that situation and all the difficult things that were happening for that family, so I inquired at the milk bank about donating. But because my son was already six months old, I couldn't donate. But they said if I decide I want to expand my family, then get in touch with them then.

So, a couple of years later, with our second child, I got in touch with them even before he was born to say I would like to donate. They said when my baby is 3 months old, and I have settled into feeding, then that would be the time we could set things up. They sent me initial forms, and I had to get blood tested. I got the midwife to sort that out for me. So, I was ready to donate when he was around 3 months old.

> I had mastitis early on, but that was sorted out, and I was exclusively breastfeeding him. I would pump in the evenings to donate.
>
> They had just set up a courier service, and a guy in his motorbike would show up, which was quite nice. I had a freezer, and I would check the temperature every day. I had a regular system.

Sinead went on to say that she could not remember when she stopped donating, but she believed she continued as long as she was feeding her second son, which was over a year. She also said near the end of the process that it was harder because her son was having less and less milk, so her supply was dropping, although she said she had generally had a very good supply with all of her children. She talked to us about how having an oversupply can have difficulties as well and that she used NHS breastfeeding support services to help her to deal with difficulties she encountered. Sinead went on to tell us that she did not donate with her third child because the pregnancy was very difficult, and she thought that her third child might need donor milk himself. However, in the end, her supply was very good, and she was able to exclusively feed him, although these stressors meant she did not feel capable of donating again. She went on to say,

> I told my neighbour that I was donating, and she was really touched. She would see them coming to get the milk, and she would say that was really great.

She spoke about how well her niece's child was doing, and how wonderful she felt it was that Sinead had chosen to donate in her niece's memory. Sinead talked about how close she felt to this neighbour, far closer than other neighbours, since they shared a lot of the "lives and difficulties" associated with infant feeding with each other.

Most bereavement donations occur when the infant passes, often a hospitalized infant whose mother may have been expressing for that infant. Unfortunately, in some cases, these infants do not survive, and it is the mother who donates in memory of the infant (Cassidy and Dykes forthcoming). This occurs more often among mothers whose infants are born prematurely, but there are also cases of mothers whose infants are born full term who unfortunately do not survive, and some of these mothers find donating comforting. Also, some of these mothers may only donate milk which had been previously collected for their infants while they were alive, but some choose to donate after a death and carefully define a set period in which they will make their donation.

All of the donor human milk bank services involved in our research actively accept donations from bereaved mothers, and if the mother is not able to go through the process for the milk to be used clinically, then the milk is automatically used for research. This means that a bereaved mother who wishes to donate is never turned away. Grief is, or course, very individual, and parental grief is fraught with particularly difficulties. The prevalence of bereavement donation is unknown, but it needs to be clear that there is no potential for additional harm associated with providing information to parents about this potential option. This is one issue which, among many others, calls for additional comparative research.

Families integrated care and comparative research

One of the hospital-based services involved in our research had a special discussion related to links with the Toronto-based family integrated care (FiCare) research trial running at the time (O'Brien et al. 2018). Part of this research is the recognition that integrating families in neonatal care increases maternal own milk production and leads to much more positive outcomes. All of our interviews illustrate the need for more extensive discussion related to how we imagine the mother and the baby and their combined journey into this world, but also their extended families, who are key support systems, as well as interactions with healthcare staff, including donor human milk banking staff. As we discussed in Chapter 1, and will discuss again in Chapter 6, this is linked to becoming a mother, or matrescence (see Provenzi, et al. 2016). Are they full term or preterm, and if preterm, how well is either the mother and/or the baby? What were the mother's original infant feeding plans, and how can we accommodate her original plans? The stories around human milk donation often involved very medicalized births, with mothers and/or infants often ending up in neonatal intensive care units. While we were conducting this research, colleagues published some comparative research on the importance of family-centred care in the neonatal intensive care units in three countries (England, Sweden and Finland) (Flacking, Thomson and Axelin 2016). As they discuss, there is wide variation cross-cultural regarding parental involvement in neonatal units around the world, and it is often linked more to maternal breastfeeding, although extending these interactions to include the entire family has been identified as being extremely beneficial to everyone involved (Flacking, Thomson and Axelin 2016). These authors are linked to the international and interdisciplinary research network entitled Separation and Closeness Experiences in the Neonatal Environment and part of comparative plans for future research on these issues. In our next chapter, we will discuss how these maternal and infant experiences are expanding around the world, including calls in some countries to expand these services. Clearly, comparative studies of these expanding services need to include and recognize mothers and their infants, as well as other family members involved in these exchanges, as well as the maternal generosity associated with these important gifts (Mauss 1925, 1954, 2002, 2016).

Notes

1 According to the *Oxford English Dictionary*, this is an American short form for prematurely born infant and dates back to the late 1920s.
2 Please note that the actual names and identifying features have been changed or omitted to ensure anonymity.
3 Allomaternal means "women other than mother" (Hewlett and Winn 2014: 200). Hewlett and Winn (2014) discuss allomaternal nursing and have argued that anthropological archives indicate this form of other mother nursing occurs in over 90 percent of societies studied.

References

Acharya, Chitra. 2018. "Human-Computer Interaction and Patient Safety," PhD Dissertation. Supervised by Prof Harold W. Thimbleby. Swansea University.

Adhisivam, B., B. Vishnu Bhat, N. Banupriya, Rachel Poorna, Nishad Plakkal, and C. Palanivel. 2017. "Impact of Human Milk Banking on Neonatal Mortality, Necrotizing

Enterocolitis, and Exclusive Breastfeeding – Experience from a Tertiary Care Teaching Hospital, South India." *The Journal of Maternal-Fetal & Neonatal Medicine.* 32 (6): 902–5.

Cassidy, Tanya M. 2012. "Mothers, Milk, and Money: Maternal Corporeal Generosity, Sociological Social Psychological Trust, and Value in Human Milk Exchange." Special Issue on Motherhood and Economics. *Journal of the Motherhood Initiative (JMI)* 3 (1): 96–111.

———. 2016. "Pumping for Preemies. Abstract for MAINN Conference." *Maternal and Child Nutrition.*

Cassidy, Tanya M. and Fiona Dykes. forthcoming. *Milk Tears: Ethnographic Extensions of Bereavement Donation.*

Becker, G. E., H. A. Smith, and F. Cooney. 2016. "Methods of Milk Expression for Lactating Women." *Cochrane Database of Systematic Reviews* (9). Art. No.: CD006170.

Bergman, Jill. 2010. *Hold Your Prem.* Cape Town: Kangaroo Mother Care.

Blesa, Manuel, Gemma Sullivan, Devasuda Anblagan, Emma J. Telford, Alan J. Quigley, Sarah A. Sparrow, Ahmed Serag, Scott I. Semple, Mark E. Bastin, and James P. Boardman. 2019. "Early Breast Milk Exposure Modifies Brain Connectivity in Preterm Infants." *NeuroImage* 184: 431.

Carroll, Katherine. 2014a. "Body Dirt or Liquid Gold? How the 'Safety' of Donated Breastmilk is Constructed for Use in Neonatal Intensive Care." *Social Studies of Science* 44 (3): 466–85.

———. 2014b. "Breastmilk Donation as Care Work" In *Ethnographies of Breastfeeding*, edited by Tanya Cassidy and Abdullahi El Tom. London: Bloomsbury.

DeMarchis, A., K. I. Ballard, K. A. Mansen and C. Engmann. 2017. "Establishing an Integrated Human Milk Banking Approach to Strengthen Newborn Care." *Journal of Perinatology* 37: 469–74.

Dempsey, E. and J. Miletin. 2010. "Banked Preterm versus Banked Term Human Milk to Promote Growth and Development in Very Low Birth Weight Infants." *Cochrane Database of Systematic Reviews* (6). Art. No.: CD007644.

Esterik, Penny and Richard A. O'Connor. 2017. *The Dance of Nurture: Negotiating Infant Feeding.* New York: Berghahn Books.

Hewlett, Barry S. and Steve Winn. 2014. "Allomaternal Nursing in Humans." *Current Anthropology* 55 (2): 200–29.

Flacking, R., G. Thomson, and A. Axelin. 2016. "Pathways to Emotional Closeness in Neonatal Units – A Cross-National Qualitative Study." *BMC Pregnancy and Childbirth* 16 (1): 170. doi:10.1186/s12884-016-0955-3

Hobbs, A. J., C. A. Mannion, S. W. McDonald, M. Brockway, and S. C. Tough. 2016. "The Impact of Caesarean Section on Breastfeeding Initiation, Duration and Difficulties in the First Four Months Postpartum." *BMC Pregnancy and Childbirth* 16: 90. http://doi.org/10.1186/s12884-016-0876-1

Israel-Ballard, K. A., C. Chantry, K. Dewey, et al. 2005. "Viral, Nutritional and Bacterial Safety of Flash-heated and Pretoria-pasteurized Breast Milk to Prevent Mother-to-Child Transmission of HIV in Resource-poor Countries: A Pilot Study." *Journal of Acquired Immune Deficiency Syndromes* 40, 38: 175–81.

Israel-Ballard, K. A., M. C. Maternowska, B. F. Abrams, et al. 2006. Acceptability of Heat-treating Breast Milk to Prevent Mother-to-Child Transmission of HIV in Zimbabwe: A Qualitative Study, *Journal of Human Lactation* 22 (1): 48–60.

Jovchelovitch, Sandra and Martin W. Bauer. 2000. *Narrative Interviewing* [online]. London: LSE Research Online. http://eprints.lse.ac.uk/2633. LSE Research Online: August 2007. Originally published as Bauer, Martin W. and G. NS Gaskell, eds. 2000. *Qualitative Researching with Text, Image and Sound: A Practical Handbook.* London, England: SAGE Publications.

Lau, C. 2015. "Development of Suck and Swallow Mechanisms in Infants." *Annals of Nutrition & Metabolism* 66 (Suppl. 5): 7–14. http://dx.doi.org.br/10.1159/000381361

Lopes, A.-A., V. Champion, and D. Mitanchez. 2018. "Nutrition of Preterm Infants and Raw Breast Milk-Acquired Cytomegalovirus Infection: French National Audit of Clinical Practices and Diagnostic Approach." *Nutrients* 10 (8): 1119. http://doi.org/10.3390/nu10081119

Maastrup, R., B. M. Hansen, H. Kronborg, S. N. Bojesen, K. Hallum, A. Frandsen, . . ., I. Hallström. 2014. "Breastfeeding Progression in Preterm Infants is Influenced by Factors in Infants, Mothers and Clinical Practice: The Results of a National Cohort Study with High Breastfeeding Initiation Rates." *PLoS ONE* 9 (9): e108208. http://doi.org/10.1371/journal.pone.0108208

Mauss, Marcel. 1925. *Essai sur le don: forme et raison de l'échange dans les sociétés archaiques* l'Année Sociologique, seconde série, 1923–24.

———. 1954. Sociologic et anthropologie, avec une introduction de Claude Lévi-Strauss, Paris, P.U. F., 4° édition augmentée.

———. 2002. *The Gift*. London: Routledge.

———. 2016. *Selected, Annotated*. Translated by Guyer, Jane I. The Gift.

Meier, P. P., A. L. Patel, and A. Esquerra-Zwiers. 2017. "Donor Human Milk Update: Evidence, Mechanisms and Priorities for Research and Practice." *The Journal of Pediatrics* 180: 15–21. http://doi.org/10.1016/j.jpeds.2016.09.027

O'Brien, K., K. Robson, M. Bracht, M. Cruz, K. Lui, R. Alvaro, et al. 2018. "Effectiveness of Family Integrated Care in Neonatal Intensive Care Units on Infant and Parent Outcomes: A Multicentre, Multinational, Cluster-Randomised Controlled Trial." *Lancet Child Adolescent Health* 2 (4): 245–54.

O'Reilly, Andrea. 2016. *Matricentric Feminism*, Theory, Activism, Practice. Toronto: Demeter Press.

O'Sullivan, E. J., S. R. Geraghty, and K. M. Rasmussen. 2018. "Informal Human Milk Sharing: A Qualitative Exploration of the Attitudes and Experiences of Mothers." *Journal of Human Lactation*. 32 (3): 416–24.

PATH. 2013. *Strengthening Human Milk Banking: A Global Implementation Framework*. Seattle: PATH Publications.

Picaud, J. C., R. Buffin, G. Gremmo-Feger, J. Rigo, G. Putet, and C. Casper. 2018. "Working Group of the French Neonatal Society on Fresh Human Milk Use in Preterm Infants. Review Concludes that Specific Recommendations are Needed to Harmonise the Provision of Fresh Mother's Milk to Their Preterm Infants." *Acta Paediatrica* 107: 1145–55.

Provenzi, Livio Barello, Serena Monica Fumagalli, Guendalina Graffigna, Ida Sirgiovanni, Mariarosaria Savarese, and Rosario Montirosso. 2016. "A Comparison of Maternal and Paternal Experiences of Becoming Parents of a Very Preterm Infant." *Journal of Obstetric, Gynecologic & Neonatal Nursing* 45 (4): 528.

Williams, Kelly. 2018. "Mum of Premature Baby So 'Overwhelmed' by Human Milk Bank Donations that Helped Daughter – She Got a Job There.' Daily Post, February 4, 2018. www.dailypo Stco.uk/news/north-wales-news/mum-baby-human-milk-bank-14217383

World Health Organization. 2018. *Implementation Guidance: Protecting, Promoting and Supporting Breastfeeding in Facilities Providing Maternity and Newborn Services – The Revised Baby-friendly Hospital Initiative*. Geneva: World Health Organization.

Zizzo, Gabrielle. 2013. "Perceptions and Negotiations of 'Failure' in an Australian Breast Milk Bank." In *Breastfeeding: Global Practices, Challenges, Maternal and Infant Health*, edited by Tanya Cassidy. New York: Nova Science Publishers.

6 Building liquid bridges

In this chapter, we consider the future of donor milk banking in an ever-changing world. Milk banking is defined by international research and inquiry, as well as by ever more diverse and unexpected efforts to define relevant good practice. Such efforts are contingent on transnational agreements and collaborations beyond the control of those who initiate them, with the result that the future expansion of donor human milk bank services can never be guaranteed. Social and medical historians may well conclude that the end of the second decade of the twenty-first century represented a defining and pivotal moment in the story of human milk donation in the UK and beyond. Social scientists may also conclude that this historical moment offered a remarkable opportunity for collaborations between specialisms that had never before seen themselves as cognate, let alone mutually reinforcing. The present moment (2018) illustrates a widely perceived need to transcend existing economic models of exchange and reciprocity, acknowledging the diverse cultural contexts for women helping children across the world.

In February 2016, as we started to collect our ethnographic data for our study, the *New York Times* published the obituary of Dana Raphael, a world famous anthropologist, who, along with Margaret Mead, established the US-based Human Lactation Center Ltd. to provide information to mothers and children around the world[1] and to study breastfeeding around the world, an initiative the *New York Times* had originally reported on in 1978 (Connell 1978). That same year, in September, Lois Arnold, one of the most widely published authors and advocates of donor human milk banking also passed away.[2] Called "the mother of modern milk banking", her colleague Karin Cadwell recalls that Arnold began her journey in the world of donor human milk banking originally as a donor in Hawaii at Hawaii Mother Milk Inc. (HMMI)[3] in 1978, a story Arnold narrates in her book *Human Milk in the NICU: Policy Into Practice* (2010, 314), where she says her sister, a paediatric nurse practitioner, said she had so much milk she could feed an army. As she discusses, she then started to help with the collection of milk and then began working in the office, eventually taking on the post of assistant administrator. In 1985, Arnold represented the HMMI at the inaugural meeting of HMBANA. She went on to say that she left the HMMI in 1987 to pursue a master's level degree in public health, which she points out was very different from her undergraduate training in zoology. In 1990, after completing her master's, Arnold was to become the first executive director of HMBANA (Arnold 2010, 352). In 1997, Arnold left her work with HMBANA and began to focus her activities on the National Commission on Donor Human Milk Banking, a special project of the American Breastfeeding Institute (ABI), a Massachusetts-based organization, which began in 2001 and whose stated mission is as follows:

> To promote, protect, and support breastfeeding in the United States through research and education; to foster research in all aspects of breastfeeding and child nutrition; to create a database and coordinate research projects on this subject throughout the United States; to conduct seminars, publish articles, conduct research, make grants, fund projects, and educate the public on the benefits of breastfeeding; to act as a consultant in the area of breastfeeding; to teach others about breastfeeding and human lactation; to be a voluntary health and welfare organization.

As we were preparing our ethical approval to conduct our ethnographic study on donor human milk banking, the ABI announced on its Facebook page that it was launching a World Breastfeeding Trends Initiative (WBTi) project.[4] It seems, however, that the links between the ABI and the WBTi have disappeared—a termination that may be related to Arnold's death, but which also seems related to the global politics of human milk research, which is often linked to the efforts of key individuals. In the world of donor human milk banking, as PATH says, there are key roles for "champions" to help with the expansion of donor human milk services, and several of the people involved in our study continue to be international champions of donor human milk banking. Recently, PATH published an important decision-making chart to protect, promote and support breastfeeding through the use of donor human milk (Brandstetter et al. 2018).

The long-time American business model (Schon 1963) of "champions" has become particularly prevalent in healthcare innovations (Shaw et al. 2012). Shaw and colleagues (2012) argue that there are two kinds of champions. The first is the so-called project champion who is linked to specific projects. The second kind of champion is linked to larger organizational changes. In the world of donor human milk banking, both forms of champions can be seen, often overlapping with national and international initiatives linked to breastfeeding, and to maternal and infant health issues. These are major topics for many organizations that in turn contribute to the international voices associated with these issues and help to constitute the overall political context of breastfeeding (Palmer 2009). These are also linked to the erosion of women's confidence in their own bodies through increasingly medicalized births (Dykes 2006), resulting in negative maternal experiences with breastfeeding and as a result the continued expansion of artificial feeding, despite the internationally recognized health benefits of breastfeeding for mothers and infants.

Community-based champions

As we have discussed throughout this book, the global expansion of donor human milk services is linked to survival rates in neonatal units around the world (Arslanoglu et al. 2013; Ultrera Torres et al. 2010). Primarily, banks are devoted to supplying milk for infants born prematurely, who continue to be the largest group receiving this service across the UK and many parts of the world (Weaver 2018), but these are not the only recipients. The four human milk services involved in our study also routinely supply other infants, such as infants who are born full term but who then need some form of surgery, in particular gut or cardiac surgery, some of whom are then less able to tolerate non-human formula and their mothers are not able to provide milk themselves. For some of these infants, supplying this milk helps to ensure that the last hours of the infant's life would be calmer for not only the infant but also for family members who are understandably under the unimaginable stress of dealing with a sick, and sometimes dying, infant. In addition, the Northwest Human Milk Bank has also participated in a community pilot project to facilitate mothers whose own supply needed to build up, with the help of NHS lactation counsellors.

Brazil is the world leader associated with donor human milk services, originally had infants whose mothers were HIV positive ranked above those born prematurely, still the second highest group of recipients in many parts of the world (Almeida and Dórea 2006). Similarly, in 2000, a unique community-based milk bank was established by Professor Anna Coutsoudis in South Africa for infants who had been abandoned or orphaned by mothers with HIV/AIDS (Coutsoudis et al. 2011). Connected to her research on the advantages of an exclusive human milk diet for infants born to mothers with HIV/AIDS (Reimers et al. 2018), the project originally received some funding from UNICEF and was called iThemba Lethu (a Zulu phrase which she has translated as "I have a destiny", although the literal translation is closer to "our hope") (Naicker et al. 2015). The South African system of donor human milk services has contributed to the international expansion, offering three models, including the community-based model of iThemba Lethu, the more traditional hospital-based models of smaller services and a so-called public-/private-based model associated with the South African Breastmilk Reserve,[5] which services 24 hospitals with a number of depots. The expansion of these milk banks across South Africa without legislative governance has led the Human Milk Banking Association of South Africa to develop guidelines,[6] but compliance with these guidelines continues to be an issue for donor human milk banks not only in South Africa but also around the world (Figure 6.1).

As we mentioned earlier, similar guidelines were established in the UK under the NICE system, and these continue to be the operations referenced, although compliance is not monitored, which is not an issue in Ireland or Scotland where a single service exists. In England, there are several banks where compliance guidelines could be an issue, although this may be part of the movement towards regionalization in England, and the larger banking services, including the Northwest Human Milk Bank in Chester, have a governing body within the NHS Health Trust.

Figure 6.1 Donor Recruitment from Ithemba Lethu, South Africa
Source: Permission has kindly been given by iThemba Lethu in South Africa.

A bank with a difference

While the MUIMME research project was in progress, a new donor human milk bank service was started in London (Weaver 2018; Shenker 2016), and in a guest blog from the UNICEF Baby Friendly Initiative, this "bank with a difference" discussed its plans for the future.[7] The inequity of access to donor milk services across England and Wales led to the establishment of this "community interest company" (CIC) with the following three aims (Weaver 2018, 3):

1   The main aim of the Hearts Milk Bank is to provide safe and assured supplies of donor milk to any hospital neonatal or paediatric unit that is unable to access it from local hospital-based milk banks.
2   A secondary aim is to promote and support human milk–based research, particularly into epigenetic studies of breast cancer but also in ethically approved areas of human milk and infant feeding. Areas for further research highlighted in both the NICE guideline in 2010 and the BAPM report in 2016 have not yet been addressed in the UK, and the Hearts Milk Bank aims to facilitate these and other studies.
3   The provision of safe, screened donor milk to mothers whose babies are not premature or unwell, which, when combined with high-quality IBCLC (International Board Certified Lactation Consultant)-led support ensures that women who do not wish to use formula, but (short or long term) need extra milk, can access human milk instead.

As we mentioned in Chapter 1, the BAPM report (2016) had representatives from two human milk banking services involved in our study, but unlike the NICE guidelines, it did not have a specific public and parental delegate. Also, as we mentioned, the BAPM group met twice in 2015, clearly when there were a lot of planned changes occurring among human milk services, particularly in England and the London area especially.

Goddard and Smith (2001) have defined equity of access in the UK NHS healthcare services as "the ability to secure a *specified* set of healthcare services, at a *specified* level of quality, subject to a *specified* maximum level of personal inconvenience and cost, while in possession of a *specified* amount of information" (p. 1151). Researchers have argued that equity of access is complicated and involves not only social constraints (such as lifestyle, age, gender and community background) but also economic and geographic aspects (Cookson et al. 2016).

Asserting the principle of equity of access is certainly not unique to the UK nor is it unique to human milk services, although the devolved nature of healthcare services in the UK means potential inequity, as Greer (2016, 16) says there has been a movement towards "four increasingly distinct health systems, in England, Northern Ireland, Scotland and Wales", resulting in different "policies, priorities and organization". As we have discussed, the single systems of milk bank services across Scotland has been a priority for the government's breastfeeding policies. Similarly, the service in Northern Ireland was also set up to support not only cross-border health cooperation with the island as a whole but also island-wide support of breastfeeding policies. The longer history of human milk services in England has meant a more divided service, although the Northwest Human Milk Bank service in Chester offers services to all of Wales and England.

Only in Scotland, at the One Bank for Scotland, is there complete access to donor human milk services, with the only hindrance being based on healthcare provider bias or some socio-cultural objection from parents of recipients, both hindrances we see throughout the service provisions across the UK. The acceptability among physicians is changing globally, as evidence increases, as has been the case with increasing acceptance among the wider public as well. During our research, however, among some physicians who saw this service as essential, there was a discussion that parental consent should not be necessary since they considered donor human milk to be a vital health provision. This issue brings up a lot of parental issues in neonatal units and is an issue one which we will not solve at this time, but parental concerns are a key issue to consider as these services expand around the world.

Vietnam gets a bank

All of the donor human milk bank services involved in our ethnographic research have been involved in the expansion of this service around the world. During our data collection, in May 2016, the Glasgow Children's Charity hosted a visit from an international delegate of Vietnamese visitors, including both healthcare staff, potential service staff, as well as government representatives. In February 2017, as we neared the end of our ethnographic data collection, the Scottish milk bank co-ordinator was asked and supported to go to Danang, Vietnam, for the opening of the first milk bank in that area—a relationship which continues to this day. As we see on the website, and as the PATH press release states,

> The first human milk bank in Vietnam opened today at the Danang Hospital for Women and Children. Supported by the Maternal and Child Health Department, Ministry of Health; the Danang Department of Health; and the international nonprofit organizations PATH and FHI 360 (through the Alive & Thrive initiative), the human milk bank will provide lifesaving donor milk to 3,000 to 4,000 at-risk infants each year.[8]

The leading clinician involved in the establishment of this human milk bank service has given a detailed TEDx talk on the establishment of the first milk bank service in Vietnam.[9] This initiative is detailed on the PATH country profile website,[10] including a 3D virtual tour of the Vietnam milk bank.[11] This tour shows three areas for women to express, an administration/meeting table and a processing room. And as we were preparing this chapter, a narrative from a donor mother was published on Alive & Thrive,[12] which includes a link to a photojournal which they titled "From Breast to Benefit: The Journey of Donor Human Milk", showing the process of the milk from the doors of the milk bank service, to the donor, to the processing room and through the pasteurization of the milk, to the babies who receive the milk in the NICU, ending with a vision of staff which says,

> The experience in Da Nang paves the way for the development of national guidelines for human milk banking, and the establishment of additional human milk banks throughout the country, with the potential to save thousands of lives and ensure healthy growth and development for infants.

And although their first concerns are national, there are also international collaborations that are being linked to these expansions and to the regulation of the expansion of donor human milk services around the world.

Shortly after the milk bank opened in Vietnam, UNICEF was involved in the closure of a commercial-based American service in Cambodia (Murdoch 2017; Jackson 2015), and almost as soon as the doors closed in Cambodia, as Jodine Chase has stated in her blog, windows were opened in Myanmar (Chase 2017). Marion Rice (2017) also reminds us that human milk is "not just a product or a resource, it is so much more than that, it is life-giving", and therefore deep emotions underpin these exchanges. Accordingly, the potential for commercial expansion and exploitation is rife, and the unregulated expansion of donated human milk across the world continues to call attention to these issues of exploitation (Cassidy et al. 2018; Dowling and Pontin 2017; Smith 2015, 2017). In the US, a number of commercial companies have arisen that have what they call "donor milk banks", where so-called donors are compensated for their donation, which has understandably generated anxieties regarding potential exploitation (Lee 2013). In India, for instance, in 2014, the expansion of milk banks was heralded as an answer to saving babies lives (Chakraborty 2014) and could have been seen

Building liquid bridges 107

as a major part of the WHO UNICEF Global Strategy for Infant and Young Child Feeding (WHO 2003), which continues to be underfunded (Gupta et al. 2012; Smith 2015). It has more recently drawn criticism about its unregulated expansion (Gupta 2017). India is also interesting because of the international collaboration with Norway, which began in 2013 and is linked to the Norwegian government and to healthcare provisions. Norway is one of a few places worldwide which provides unpasteurized human milk to infants in need (Grøvslien and Grønn 2009).

As we have discussed throughout this book, the uneasy relationship between feminism and breastfeeding continues to be problematic. Breastfeeding continues to be a maternal link to inequitable parenting, but as Lee (2016) points out, regardless of sex or gender, it is incumbent on us to "feed the other regardless of whether we have given birth to her". The patriarchal privatization of the maternal-infant dyad is what leads to these problems, whereas notions of maternal generosity and embodied exchange offers potential emancipation (Cassidy 2012).

Medicalized milk sharing in Malaysia

During our ethnographic data collection, a team of interested individuals from Malaysia visited the Northwest Human Milk Bank in Chester, leaving a framed picture of the house as a gift (Figure 6.2).

Expansion of donor human milk services across the Islamic world has been severely hindered (Thorley 2016; AlHreashy 2018). The complexity of Islamic theology and teaching further complicates such issues, which are often variant according to cultural interpretations (Clarke 2007; Cassidy and El Tom 2010). In 2011, the 97th Muzakarah of National Fatwa Council for Islamic Affairs Committee the prohibited implementation of a milk bank because of "the possibility of

Figure 6.2 MUIMME fieldnotes from the Northwest Human Milk Bank, Chester, UK and taken by Tanya Cassidy.

overlapping of the progeny (nasab)" (Daud et al. 2016, 509). In the US, because milk is pooled between mothers, this is seen to be incompatible with Islamic milk kinship (EL-Khuffash and Unger 2012). But in the UK, human milk is never pooled between mothers, but is instead pooled, as we mentioned earlier, from several expressions from one mother, which is directly connected to the traceability of the milk back to specific women, and the NICE guidelines (2010) specifically state that records must be kept for 30 years, both of which should indicate that this service is potentially compliant with Islamic milk kinship laws (Williams et al. 2016).

At the beginning of our research (in 2015), a meeting to discuss issues related to Islam and human milk donation was convened between interested stakeholders, including religious leaders, healthcare providers, a representative of UKAMB and an anthropologist who has worked on Islamic milk kinship issues for several years (Williams et al. 2016). Although this was published in a prestigious medical journal, with a statement available on the UKAMB website, milk banks which deal with either donors or parents of recipients who follow the Islamic faith continue to run into difficulties of interpretation based on misunderstandings, and so this is an area that needs additional research in the future.

However, alternative interpretations may exist in other countries (Khalil et al. 2016). In 2017, the Islamic Religious Council of Singapore said that the extreme medical need of premature infants means that the fatwa is lifted and that these infants can and should receive donated human milk if their own mothers are not able to provider their own milk for some reason (Muis 2017). However, despite the recognition from religious leaders of the necessity of donor human milk, in particular for prematurely born infants, there continues to be potential and actual reluctance on the part of the public in terms of acceptance of donor human milk, which has also formed part of the problem with the establishment of any donor human milk banks in Turkey where it was found (Senol and Aslan 2017) that although over half of the women they surveyed would have considered being a donor, fears about transmission of disease and religious barriers were still a hindrance.

Brexit and the future

In the UK, and particularly on the island of Ireland, the future of milk banking is further complicated by Britain planning to leave the EU (so-called Brexit—British Exit). As we are writing, the UK is undergoing the controversial and convoluted process of leaving the EU, having invoked Article 50 of the EU constitution. This follows a referendum on EU membership held in June of 2016.

The Scottish milk banking services may remain comparatively unaffected by Brexit insofar as Scotland has a single bank in Glasgow which serves Scotland as a whole. The case of Northern Ireland, however, is considerably more complex. The community milk bank in Co. Fermanagh serves the whole island of Ireland, receiving milk donations from across the island and distributing milk across the island on the basis of need. Needless to say, this process has been enabled by an invisible (so-called frictionless) border between Northern Ireland and the Republic of Ireland that was made possible by the Good Friday Agreement signed in 1998.

This agreement, which has substantially demilitarized the region, was entered into on the basis that Ireland and the UK, as EU members, had no trade barriers and therefore no reason to impose significant border controls. The retention of this critical aspect of the Good Friday Agreement in the case of one party leaving the EU was not the subject of high profile or decisive debate prior to the referendum. While all parties to the Brexit process profess their commitment to what is termed a "fictionless" border on the island of Ireland, the practical means of so doing appear to be elusive. What is (at time of writing) an invisible border separating different Irish counties belonging to separate polities will soon become the border between an EU common customs and travel union and a UK determined to remain outside such arrangements.

If any version of a hard border and/or a heavily policed border between Northern Ireland and the Republic of Ireland emerges, then island-wide, cross-border health cooperative human milk bank service will undoubtedly suffer. In many ways, meanwhile, this border question provokes more widely applicable questions regarding the geographical reach of individual human milk bank services and the need to institute a degree of flexibility when it comes to supplying infants in need. In January 2018, the bank in Northern Ireland finally moved from its community premises to a larger, more sterile hospital facility in Enniskillen Acute Hospital—a move which had been planned during our data collection, but which took some time to realize. The move to a hospital setting, however, does not negate the community aspiration which the staff running the milk bank service still intends to uphold. Much like the Scottish wide service which is also located in hospital, this relocation is more about infrastructure and governance than service provision per se. At the time when the move occurred, there was a lack of milk for some of the hospitals on the island of Ireland, and the Northwest Human Milk Bank was able to ferry supplies across the Irish Sea, another international cooperation for sending milk to the Coombe hospital in Dublin. Blood bikers transported the milk to the ferry, where it was stored in a freezer and then picked up by volunteer bikers again on the Irish side to take the milk to the hospital. But despite this cooperation, certainly the Northwest Human Milk Bank was not the closest bank to this infant need; it was, however, the closest with a surplus supply and the ability to transport it to child in need.

An important point to recall is that the Milk Bank in Northern Ireland is also connected to the expansion of milk banking services across Europe, in particular, when in 2007, after the tenth anniversary of UKAMB, Aleksandra Wesolowska from Poland visited the bank in Northern Ireland and then used what she learned to help to re-establish milk banking services in Poland by opening one in Warsaw. This bank has also integrated with the expansion of breastfeeding in Poland and uses the following "Milky Way" poster for donors and recipients (Figure 6.3).

Aleksandra has a doctorate in biology and linked the bank she established in Warsaw to research, but also to supporting mothers in feeding their own babies, while encouraging those who are able to donate to help others. Recently, the Polish Ministry of Science gave her an award for her work on human milk and lactation. Her efforts have seen donor human milk services expand across Poland, where in January 2019, there are 11 active services across Poland and 3 more being planned.

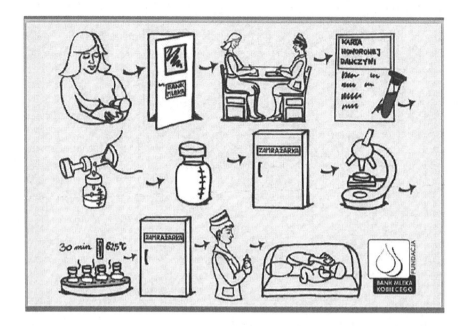

Figure 6.3 A Polish Picture of the Process of Donor Human Milk Banking
Source: Reprinted with kind permission from Aleksandra Wesolowska.

If the Republic of Ireland were to acquire a milk bank of its own, then the border question would remain pertinent, since many border counties within the Republic remain much closer to Fermanagh than to Dublin (or wherever such a bank might be). Any imposition of traditional border controls, therefore, potentially jeopardizes the life chances of vulnerable neonates and threatens a system of exchange that depends on goodwill, flexibility and an ability to improvise.

Human milk research and the future

In the (likely) event of a milk bank establishing itself in the Republic of Ireland, it would be most useful to designate such a facility as a "research bank". There are many European and other international examples for such a bank, which are invested in clinical and research applications of human milk to differing extents. Our research suggests that we could link for "ideal types" of donor human milk services based on links to research. The first we could call a clinical bank with few or no links to a research laboratory. Such banks typify a certain efficient simplicity demonstrated by the quoted phrase "it's not rocket science". Such banks are likely to stress the obviousness of human milk as a "natural" choice and pre-eminence of clinical need, while nonetheless adhering to the most stringent and scientifically demonstrated standards of hygiene and safety. Several of the smaller milk banks in England would fit into this category, especially those that only service their own units, although it is not unheard of for these units to participate in research, it is just much less a priority.

A modification of this first category would involve a clinical bank in which a few bio-samples are regularly, but not routinely, stored for possible research. This second type describes clinical and research banks in which milk research is a normal and expected function of the bank, but the day-to-day operation of the bank is dictated by clinical concerns. A key feature of this type of bank is that only milk which cannot be used clinically can be directed towards research. Only when clinical demands are more or less satisfied can research be either initiated or developed. These two types, in fact, capture the features of the four main milk banks involved in our ethnographic research. All four banks are involved in research, although for the community-based banks, research is not as key a focus, although they do participate in it and are extremely helpful, their main focus is clinical and supplying donor human milk for clinical uses.

A third type of donor human milk service in which research imperatives define the organization of the bank also routinely supplies donor milk for clinical purposes. The hospital-based services in both London and Glasgow could be argued to be of this type, especially because they are both linked to university teaching hospitals, and therefore research is key, although the clinical obligations remain crucial. Another interesting example of this type of bank can be found in the US and is linked to a controversial commercial human milk company.[13] The feature of whether these services should be allowed to use the terms "donor" is debatable since they involve commercial exchange. The Northwest Human Milk Bank Service which sells directly to service uses (whether hospitals or individuals) is keen to stress that it is not charging for the milk, since this has been donated free, but instead covering its processing costs.

The fourth type of donor milk service linked to research is one wholly devoted to research with little or no clinical links. Currently, there is no bank in the UK which fits this type. However, in the US, there is such a bank type represented by the Mommy's Milk Human Milk Biorepository at the UCSD, which seems to be linked to some very interesting and significant breastmilk scientific research. Another complicated example can be found through an analysis of the albeit not complete online Global Biobank Directory, Tissue Banks and Biorepositories which lists the Coreva Human Milk Bank in Westlake Village, California, as the only explicitly stated biobank with human or breastmilk. However, the hyperlink sends you to a web page that says the account is suspended for nationalmilkbank.org. If we looked at the archives web pages for nationalmilkbank.org, we see that the NMB was started in 2005, and it is "the nation's first virtual human milk donation organization" and is linked to the commercial company Prolacta Bioscience,[14] which dates back to 2001 and is then again linked to this commercial company and not to a donor service.

In addition to the invaluable immunological research that is enabled by a research bank, a research bank has the additional advantage that it can authentically claim that "not a drop is wasted". Instead of overburdening donors and potential donors with questionnaires and other forms of interrogation associated with health and lifestyle, donors can be informed from the outset that any milk that is tested and found unsuitable for clinical application will serve a research purpose that will in the future help develop clinical applications. From a research point of view, there can be no such thing as "bad milk", and therefore many of the more sensitive issues of "rejection" associated with milk donation can be alleviated if not eliminated.

As this book has demonstrated, donating milk is unlike donating money or other forms of personal property: It is a donation from the self and of the self, and the suggestion that such a donation might be unwanted, unusable, wasted or "bad" is therefore especially urgent and concerning. From an immunological point of view, meanwhile, securing a continuous supply of a substance rich in genetic cellular possibility is a particular defining need. Establishing a research bank that accommodates laboratory analysis alongside direct neonatal provision requires a unique form of partnership between the social sciences and the so-called hard sciences—a recognition of an exciting, transformative and potentially lifesaving relationship of symbiotic co-dependency. In the (likely) event of a milk bank establishing itself in the Republic of Ireland, it would be most useful to designate such a facility as a "research bank". There are many European and other international examples for such a bank which are invested in clinical and research applications of human milk to differing extents.

Such is the state of indeterminacy surrounding the state of UK and global milk banking, that composing a "conclusion" to this chapter, let alone the book as a whole, is a somewhat tendentious exercise. Subject to various political and economic pressures, as well as alert to political and economic opportunities, milk banking may expand or contract in ways that its champions and advocates are incapable of either foreseeing or forestalling. What can be asserted far more confidently is that the knowledge base on which milk banking policy depends will become more exhaustive and more interdisciplinary. Milk banking may or may not be "rocket science". Milk banking may regard itself as a bio-technology or merely as the streamlined expression of an ancestral sense of feminized community. Above all, milk banking depends for its promotion on the use of personal testimonies and the sharing not merely of milk but also of narratives. From an anthropological point of view, the story of the development of donor milk provision illustrates a kind of courageous sharing of the self for a common good that is both powerfully communitarian and perilously intimate and individual at one and the same time. The ongoing theorization of this movement has powerful implications for a feminist philosophy of selfhood and othering in the twenty-first century.

Notes

1 See www.thehumanlactationcenter.com.
2 See www.ourmilkyway.org/remembering-lois-d-w-arnold-phd-mph-alc/.
3 This service started primarily as a milk bank in 1975, although it no longer is a bank service. It has evolved as an extensive breastfeeding support service in Hawaii, although they may be re-establishing their milk banking procedures in the near future. See http://himothersmilk.org.

4 See www.facebook.com/American-Breastfeeding-Institute post 8 August 2015.
5 See www.sabr.org.za/.
6 See www.hmbasa.org.za/about-us/history/.
7 See www.unicef.org.uk/babyfriendly/hearts-milk-bank-bank-difference/.
8 See www.path.org/news/press-room/800/.
9 See www.youtube.com/watch?v=hjE2cfpdsYo.
10 See http://sites.path.org/vietnam/reproductive-maternal-and-child-health/human-milk-banking/.
11 The 3D tour can be accessed by going to http://venue.rsdesigns.xyz/3d-model/human-milk-bank-da-nang/.
12 See www.aliveandthrive.org/resources/with-her-own-newborn-twins-in-the-neonatal-intensive-care-unit-this-first-time-mother-is-helping-other-newborns-thrive/. "Alive & Thrive is managed by FHI 360 with funding from the Bill & Melinda Gates Foundation and other donors" (www.aliveandthrive.org/about-us/). FHI 360 is a registered trade mark of Family Health International (www.fhi360.org/about-us).
13 In June 2017, Arun Gupta, a leader in breastfeeding research from India, discussed potential commercial exploitation of human milk banks on BabyMilkAction.org.
14 See www.prolacta.com.

References

AlHreashy, F. A. 2018. "Non-Maternal Nursing in the Muslim Community: A Health Perspective Review." *Journal of Clinical Neonatology* 7: 191–97.

Almeida, S. G. and J. G. Dórea. 2006. "Quality Control of Banked Milk in Brasília, Brazil." *Journal of Human Lactation* 22: 335–39.

Arslanoglu, S., G. E. Moro, R. Beliu, et al. 2013. "Presence of a Human Milk Bank is Associated with Elevated Rate of Exclusive Breastfeeding in VLBW Infants." *Journal of Perinatal Medicine* 41: 129–31.

Arnold, Lois D. W. 2010. *Human Milk in the NICU: Policy into Practice*. Boston: Jones and Bartlett Publishers.

Brandstetter, Shelley, Kimberly Mansen, Alessandra DeMarchis, Nga Nguyen Quyhn, Cyril Engmann, and Kiersten Israel-Ballard. 2018. "A Decision Tree for Donor Human Milk: An Example Tool to Protect, Promote, and Support Breastfeeding." *Frontiers in Pediatrics*. www.frontiersin.org/article/10.3389/fped.2018.00324

Chakraborty, Shruti. 2014. "The Answer to Saving India's Babies is as Old as Humankind." Quartz, India. https://qz.com/1292598/why-a-good-monsoon-wont-make-food-any-cheaper-in-india/

Cassidy, T. M., S. Dowling, B. P. Mahon, and F. C. Dykes. 2018. "Exchanging Breastmilk: Introduction." *Maternal and Child Nutrition* 14 (S6): e12748

Chase, Jodine. 2017. "*Human Milk News*." http://bfnews.blogspot.com/2017/11/when-doors-close-in-cambodia-windows.html

Clarke, Morgan. 2007. "The Modernity of Milk Kinship." *Social Anthropology* 15: 287–304.

Connell, Lise. 1978. "Westport Center Studies Breastfeeding Around the World." *New York Times (1923-Current file)*, February 26. Accessed May 31, 2018. http://ezproxy.uwindsor.ca.ledproxy2.uwindsor.ca/login?url=https://search-proquest-com.ledproxy2.uwindsor.ca/docview/123803450?accountid=14789.

Cookson, R. A., C. Propper, M. Asaria, and R. Raine. 2016. *Socioeconomic Inequalities in Health Care in England*. (CHE Research Paper; No. 129), 1–34. York, UK: Centre for Health Economics, University of York.

Coutsoudis, Irene, Miriam Adhikari, Nadia Nair, and Anna Coutsoudis. 2011. "Feasibility and Safety of Setting Up a Donor Breastmilk Bank in a Neonatal Prem Unit in a

Resource-Limited Setting: An Observational, Longitudinal Cohort Study." *BMC Public Health* 11: 356.

Daud, Normadiah, Nadhirah, Nordin, Zurita Mohd Yusoff, and Rahimah Embong. 2016. Chapter 46 "The Development of Milk Bank According to Islamic Law for Preserving the Progeny of Baby." In *Contemporary Issues and Development in the Global Halal Industry*, edited by Siti Khadijah Ab. Manan, Fadilah Abd Rahman, and Mardhiyyah Sahri. Springer: Syngapore.

Dowling, S. and D. Pontin. 2017. "Using Liminality to Understand Mothers' Experiences of Long-term Breastfeeding: 'Betwixt and Between', and 'Matter Out of Place'." *Health: An Interdisciplinary Journal for the Social Study of Health and Illness* 21 (1): 57–75.

Dykes, Fiona. 2006. *Breastfeeding in Hospital Mothers, Midwives and the Production Line*. London: Routledge.

EL-Khuffash, A. and S. Unger. 2012. "The Concept of Milk Kinship in Islam: Issues Raised When Offering Preterm Infants of Muslim Families Donor Human Milk." *Journal of Human Lactation* 28 (2): 125–27.

Goddard, Maria K., Paul V. Croke, and Louis B. Smith. 2001. "Equity of Access to Health Care Services: Theory and Evidence from the UK." *Social Science & Medicine* 53 (9): 1149–62.

Greer, S. L. 2016. "Devolution and Health in the UK: Policy and Its Lessons Since 1998." *British Medical Bulletin* 118 (1): 16–24. http://doi.org/10.1093/bmb/ldw013

Grøvslien, A. H. and M. Grønn. 2009. "Donor Milk Banking and Breastfeeding in Norway." *Journal of Human Lactation* 25: 206–10.

Gupta, Arun. 2017. "India Is Planning a Network of Human Breast-Milk Banks but First Needs to Put Safeguards in Place." *Scroll*, June 21. https://amp.scroll.in/article/841072/india-is-planning-a-network-of-human-breast-milk-banks-but-first-needs-to-put-safeguards-in-place

Gupta, A., R. Holla, J. Dadhich, and B. Bhatt. 2012. *The World Breastfeeding Trends Initiative (WBTi). Are Our Babies Falling Through the Gaps? The State of Policies and Programme Implementation of the Global Strategy for Infant and Young Child Feeding in 51 Countries*. New Delhi, India: Breastfeeding Promotion Network of India (BPNI)/International Baby Food Action Network (IBFAN)-Asia.

Jackson, Will. 2015. "Local Breast Milk for Sale in the US." The Phnom Penh Post.|www.phnompenhpost.com/national/local-breast-milk-sale-us

Khalil, A., R. Buffin, D. Sanlaville and J. C. Picaud. 2016. "Milk Kinship is Not an Obstacle to Using Donor Human Milk to Feed Preterm Infants in Muslim Countries." Acta Pædiatrica 105: 462–67.

Lee, Robyn. 2013. "Breastmilk Exchange and New Forms of Social Relations." *MP: An Online Feminist Journal* 4 (1): 36. http://academinist.org/wp-content/uploads/2013/05/03_MP_SPRING_Lee_Breastmilk.pdf

———. 2016. "Feeding the Hungry Other: Levinas, Breastfeeding, and the Politics of Hunger." *Hypatia: A Journal of Feminist Philosophy*. 31 (2): 259–74.

Muis. 2017. "Administration of Muslim Law Act (Chapter 3, Section 32) Fatwa Issued by Fatwa Committee, Islamic Religious Council of Singapore." www.muis.gov.sg/-/media/Files/OOM/Fatwa/Fatwa-Text-on-Milk-Bank-English.pdf

Murdoch, Lindsay. 2017. "Cambodia's Ban on Breast Milk Sales Throws Spotlight on Growing International Trade." The Sydney Morning Herald. www.smh.com.au/world/cambodias-ban-on-breast-milk-sales-throws-spotlight-on-growing-international-trade-20170329-gv8z5u.html

Murphy, J., M. E. Sherman, E. P. Browne, A. I. Caballero, E. C. Punska, R. M. Pfeiffer, ... K. F. Arcaro. 2016. "Potential of Breastmilk Analysis to Inform Early Events in Breast

Carcinogenesis: Rationale and Considerations." *Breast Cancer Research and Treatment* 157 (1): 13–22.

Naicker, M., A. Coutsoudis, K. Israel-Ballard, R. Chaudhri, N. Perin, and K. Mlisana. 2015. "Demonstrating the Efficacy of the FoneAstra Pasteurization Monitor for Human Milk Pasteurization in Resource-Limited Settings." *Breastfeeding Medicine* 10 (2): 107–12.

Palmer, Gabrielle. 2009. *The Politics of Breastfeeding*. London: Pinter and Martin, Ltd.

Reimers, Penelope, Natalie S. Shenker, Gillian Weaver, and Anna Coutsoudis. 2018. "Using Donor Human Milk to Feed Vulnerable Term Infants: A Case Series in KwaZulu Natal, South Africa." *International Breastfeeding Journal* 13: 43.

Rice, Marion. 2017. "Biological Integrity-Ethics and Control Over *Human Milk*." *Gold Lactation*. www.goldlactation.com/

Schon, D. A. 1963. "Champions for Radical New Inventions." *Harvard Business Review* 41: 77–86.

Shaw, E. K., J. Howard, D. R. West, B. F. Crabtree, D. E. Nease, Jr, B. Tutt, and P. A. Nutting. 2012. "The role of the champion in primary care change efforts: from the State Networks of Colorado Ambulatory Practices and Partners (SNOCAP)." *Journal of the American Board of Family Medicine: JABFM* 25 (5): 676–85.

Senol, Derya Kaya and Ergul Aslan. 2017. "Women's Opinions About Human Milk Donation and Human Milk Banking." *Research Article – Biomedical Research* 28 (15).

Shenker, Natalie Susannah. 2016. "The Hearts Milk Bank – The Bank with a Difference." Guest blog. www.unicef.org.uk/babyfriendly/hearts-milk-bank-bank-difference/

Smith, Julie. 2015. "Markets, Breastfeeding and Trade in Mothers' Milk." *International Breastfeeding Journal* 10: 9. https://doi.org/10.1186/s13006-015-0034-9

———. 2017. "Without Better Regulation, the Global Market for Breast Milk will Exploit Mothers." *The Conversation*.

Thorley, Virginia. 2016. "Milk Kinship and Implications for Human Milk Banking: A review." *Women's Health Bulletin* 3: 3.

Utrera Torres, M. I., C. Medina Lopez, S. Vazquez Roman, et al. 2010. "Does Opening a Milk Bank in a Neonatal Unit Change Infant Feeding Practices? A Before and After Study." *International Breastfeeding Journal* 5: 4.

Weaver, Gillian. 2018. "Banking on Change at Hearts." *AIMS Journal* 30 (3), ISSN 2516-5852. www.aims.org.uk/pdfs/journal/732

WHO (World Health Organization). 2003. *Global Strategy for Infant and Young Child Feeding*. Geneva: World Health Organization.

Williams, T. C., M. Z. Butt, S. M. Mohinuddin, A. L. Ogilvy-Stuart, M. Clarke, G. A. Weaver, et al. 2016. "Donor Human Milk for Muslim Infants in the UK." *Archives of Disease in Childhood. Fetal and Neonatal Edition* 101: F484–F485.

Endword

On the first day of my scheduled field observational research in my first milk bank, I had an interview with one of the staff members who was to become my key contact at that bank, and I asked her to tell me her story about milk banking. Frances told me about the first phone call she took shortly after starting to work in the milk bank. She said the mother sounded happy and was telling her about how she was interested in donating her milk to help vulnerable infants. The staff member said she was very happy for her to donate, but that she would need to go through a verbal health screening, and later she would need to have a blood test as well to make sure that her milk could be used clinically for vulnerable babies, but if not, she could donate for research, all of which the mother readily agreed. And so the staff member began the screening with what was considered a simple question: How old is your baby? At that point, the mother broke down crying and told the staff member that her baby had died. They both began to cry, and they eventually were able to complete the questionnaire, and the mother was able to donate her milk.

The last day of my scheduled observational research in the field, I was at one of the other milk banks, and the staff members were all busy with things, and so I offered to help by answering the telephone so that they could all do other necessary work. The telephone rang, and I was polite and began to ask how we might be able to help. The mother told me that she wanted to donate and that her baby had died, but that she was told she could still donate. I told her how sorry I was to hear about her baby and that I was helping staff members, who were far better equipped to explain to her about donation. One of the main managers took the telephone call, and with the skills gained from years of dealing with mothers who have experienced the pain of bereavement, she talked to this mother, who also made a donation.

Marilyn Strathern (1987) refers to "*anthropology* at home" as "auto-*anthropology*", unlike the more traditional presentations offered by others (Anderson 2006; Denzin 2014; Ellis 1997; Ellis and Bochner 2000, 1996; Ellis, Adams and Bochner 2010; Reed-Danahay 1997). Shortly after starting my EU fellowship, I gave a workshop on auto-ethnography, which was attended by staff and post-graduate students interested in the subject at UCLan. But this is not an ethnography about me, but about the relations surrounding donor human milk, and those relations involve me, but also my co-authors, and my story is only one part of these relations. My field notes are from the first day that I presented my own story about milk banking to the tenth anniversary celebrations of UKAMB, and although the staff member had not attended the celebrations, she was aware of my personal story, as were all of the members of the staff involved in our research. And my notes from the last day and the undirected links of having to deal again with a bereaved mother indicated to me that I needed to remember that this was an issue that permeates milk banking and the gift itself, so our first paper from that data was about bereavement and donation (Cassidy and Dykes, forthcoming). And although I mention my earlier publications, which detail my own experiences throughout this book (Cassidy and El Tom 2010; Cassidy and Brunström 2015), and we discuss bereavement and donation in Chapter 5, I wanted to wait to discuss my own experiences until the end of this ethnography. I did, however, in the preface offer the reader the opportunity to make their own choice and to decide to read a version of my story first.

My personal experiences have shaped my understanding and applied considerations for milk banking, but I am also aware that my story is very emotional and that some healthcare providers are not comfortable with recognizing the role of reflexivity in research, a key feature for feminist ethnographers, so I have decided that I would wait to tell my story at the end of this ethnography. Not because I don't want the reader to know my story, which I have referenced throughout this narrative, but because I hope the reader recognizes that my story is not the only frame around this ethnography, which is linked to the collaborative nature of the narrative as a whole. But this part of our discussion is my story and is therefore only a collaboration with my children and my partner with whom I conceived them, all of whom I asked to read this narrative before publication.

I have told my story in person and in emails to many people, but the first time I told my story in public was in 2007, before I had ever published it, at the tenth anniversary of UKAMB. I began by saying that I wore many hats and told people that I was not only a Chicago-trained social scientist who was born in Canada and who had lived and worked in Ireland for many years, but I am also the mother of two preemies, and so I wanted to tell my story again; therefore, I was publishing for the first time the narrative I told at the beginning of my journey into the world of milk banking.

Both of my sons were born at approximately 30 weeks gestation, just under a year apart, what I have been told here in Ireland are called Irish twins, but they were born an ocean apart, and when it comes to milk banking a world apart.

Liam, my first son, was born 24 February 2005, in Windsor, Ontario Canada. Although I was born in Ontario, and all of my family lives in the Windsor area, I had lived just outside of Dublin, Ireland, for over a decade. My partner and I had taken an opportunity to come and teach at the University of Windsor for a year and a half, when unexpectedly, but very pleasantly, I discovered that I was pregnant. Approximately a month after this, I went into premature labour. My partner was back in Ireland, and I was alone, so I had to drive myself to the hospital in Windsor, and there I was delivered with an emergency caesarean, as my son was born a footling breech with a true knot in his cord. My partner arrived from Europe within less than 24 hours to see both Liam and I recovering well.

Liam was wonderful, but small. He thrived for a time, but got an infection, from which he recovered but was weakened. After 16 days of life, he was diagnosed with NEC. He was rushed to the Children's Hospital of Michigan in Detroit, one of the larger and best-equipped neonatal units in the US, just across the border from Windsor. (They actually closed down the tunnel between the two countries so Liam's ambulance could get through as quickly as possible. A little-known arrangement which has existed for a number of years between the two hospitals in order to care for the sickest of babies.) After confirming the diagnosis, they operated on him but immediately the neonatal surgeon decided that there was not enough bowel left to be "compatible with life". Several hours later, Liam died in my arms.

Throughout Liam's short life, I expressed and tried to produce milk for him, but, unfortunately, my caesarean scar became infected, and I was never able to produce much milk at all for him. The hospital neonatologists, however, aggressively gave him formula. At the time, I was unaware of the life-threatening implications formula posed for preemies. After Liam died, I, as an academic researcher, wanted to understand what happened and quickly discovered the research related to NEC and breastmilk. I am a medical social scientist and therefore had experience with the literature on medicalized birthing issues and had at my disposal the tools and ability to begin to try to understand what happened to my son.

I then asked the question about what happens when women are not able to produce milk for their prematurely born infants and was drawn into looking at milk banks. At the time, there was only one milk bank in Canada, thousands of miles from Windsor, so Liam was never going to get milk from a bank. I then discovered that there was also only one milk bank in Ireland, albeit in Northern Ireland and therefore under the control of the UK Department of Health. It does, and did, however, provide milk for any baby in need on the island as a whole.

I was in contact with the manager at the Northern Ireland milk bank and began developing ideas associated with a larger research project on milk banking when we were blessed with another pregnancy in the summer of 2005, which has resulted in my beautiful, thriving son Gabriel. I used the knowledge I had gathered to determine what I believed was the best possible birthplace for my second son.

My partner and I determined that the medical facilities available at the high-risk maternity hospital in Dublin offered the highest level of care both in skill and technology, and would be comparable to the best units in major urban centres. Unlike Toronto or Montreal in Canada, however, the availability and willingness to use a milk bank made Dublin a better place for me to give birth.

My pregnancy with Gabriel was complicated by complete placenta previa, which resulted in a number of bleeds and a long hospitalization on my part. I made it to 30 weeks plus 3 days gestation when I bled so severely that the baby was in distress, and I needed to be delivered once again with an emergency caesarean.

Gabriel was born in the small hours of the morning of 15 February 2006. I was transfused with five units of blood and two units of plasma, and my son was immediately put in a NICU, where he also received a transfusion.

My partner talked to Gabriel's neonatologist about getting donor milk, and the milk was present the next day. My partner also arranged for me to have a hospital-grade pump in my room (which involved having to go and rent one on the other side of Dublin), so I started pumping as soon as I was physically able (which, unfortunately, was not until the next day). However, due to Gabriel's health, he developed a collapsed lung. They chose not to feed him right away, and I managed to make a few mils and Gabriel's first feeds, for about three days, were exclusively my milk, but then his demand outgrew my supply, and we turn to the donor milk mixed with my milk.

Gabriel spent two weeks in NICU and then three weeks in the less intensive baby unit. He was gavage fed for most of this time, and I contacted a lactation consultant to try to get him ultimately to breast and to help increase my supply. Despite the wonderful technology and skills of many of the midwives and physicians in the hospital, there was a real push to bottle feed Gabriel, not formula, but the combination of donor milk and my own breastmilk. Due to being quarantined in the hospital to keep out a winter flu bug, my lactation consultant was not allowed access to the baby unit. This first picture is the one which began Gillian's lovely song.

After he was 3 weeks old, I again haemorrhaged quite badly (needing 10 units of blood transfused this time), and in order to save my life, I needed an emergency hysterectomy. I was so determined to make milk for my son, the day after I had this surgery, while still in high-dependency care, I expressed milk (about 70 mls), but the donor milk was his main source of nutrition.

The great news was that throughout his stay in hospital, Gabriel did not develop any infections whatsoever. When he was 5 weeks old (35 weeks gestation) and 5 pounds weight, we were allowed to take him home. The donor milk continued to be supplied at home, and I continued to express, and I was able to have the attention of my lactation consultant, who was an incredible support, although my low supply problems continued, and I never was able to obtain a full supply.

Gabriel was exclusively fed donor milk and my expressed mother's milk until he was about 15 weeks actual age, or approximately one month corrected.[1] At which time, he weighed over 11 pounds and was thriving. Unfortunately, due to shortages in the donor milk bank, and the needs of sicker babies, Gabriel was taken off the list for donor milk, and we had to put him on formula, although he was supplemented with my expressed milk. We were and still are so appreciative of the lifesaving donor milk Gabriel received; we have actively participated in a number of media discussions on milk banking (including newspapers, television and on the Internet).

At the time I first gave the talk, Gabriel was 6 years old, and at the time of this writing, he is almost 13 years old. I expressed small amounts of mother's milk for Gabriel each day until he was 2 years old corrected, which I linked to the long culturally defined rules associated with milk kinship, which as a cross-culturally orientated social scientist, and in particular, thanks to my research colleague, I looked to the Koran and the Muslim tradition where wet nursing is explicitly described and prescribed in medically related cases, and breastfeeding is generally encouraged to age 2 (Cassidy and El Tom 2010).

As an aside to the work I have been conducting on the topic of human milk banking, I thought I might mention the extended connections Gabriel's experiences have resulted in. Specifically, Gabriel's neonatologist is Eugene Dempsey, the main co-author of the most recent 2010 *Cochrane Review* comparing banked preterm versus banked term human milk to promote growth and development in very low birth weight infants. This is particularly relevant in Ireland, where a number of neonatologists were negative about donor human milk banking because it comes from a community-based bank where the majority of donor milk is from full-term babies. There are a lot of benefits to the community-based bank, and Gabriel is a key case where he gained weight at a wonderful rate. But this is linked to Gabriel's main donor who gave birth around the same time to a very large 13-pound baby, whom I was told at 1-month-old was over 20 pounds. This mother's milk made both her little person and mine super healthy and heavy. Unfortunately, this is not the case in a number of hospital-based banks and continues to be one of the main reasons why donor milk is not supported by all neonatologists.

Ireland has been a strong supporter of the Cochrane collaboration, and I originally received a Cochrane fellowship to conduct a review linking my previous work on alcohol with my current work on breastmilk with an Australian colleague, Roslyn Giglia (Cassidy et al, forthcoming). Our plan is to then extend this, as was suggested by an Ecuadorian neonatologist and director of an Ecuadorian milk bank who heard me speak about my Cochrane work and encouraged me to extend my research to donor human milk banking, which eventually became our successful MSCA fellowship and the basis for this book.

Becoming a parent is a major life-changing experience, and for those who give birth prematurely, these events often mean that your child may need medical care. As we have discussed, premature birth is an increasing global health concern, and one of the leading causes of death among these often tiny neonates is NEC, which research indicates is significantly reduced with an exclusive maternal milk diet, making this more medicine than nutrition. But also, as we have discussed, a significant percentage of mothers who give birth early experience delayed lactogenesis and or low milk supplies. And although MOM is always the best, historically and cross-culturally, many neonatal units use preterm formula, increasing the potential for poorer outcomes. But for over a century, a "low-tech" intervention of donor human milk banking (which we also noted in Chapter 2 were originally called directory or bureau) was used to offer human milk to these vulnerable infants. Growing research, including a recently published study (Adhisivam et al. 2017) from India, shows that the introduction of a donor human milk bank not only reduces morbidity and mortality but also significantly increases the percentage of infants who continue to be exclusively breastfeed for longer periods of time—a health policy for many governments around the world.

Ethnography, an important anthropological method and an increasingly important form of research for health studies, helps reveal the unspoken social and cultural patterns that shape behaviours, offering complex, detailed information for particular settings, in this case the medically controlled organization of mothers' milk. It has been 21 years since I completed my doctoral training at the University of Chicago, where I was encouraged to recognize the value of reflexive research on personal topics. Bringing one's talents to bear on a topic close to one's heart means that you have the potential to contribute in ways that others may not. As a socio-cultural medical researcher, I strive to understand not only diseases, treatments and preventions but also public and patient experiences throughout history and across cultures.

I have constructed myself as a modern global citizen. I was born and grew up in Canada with family narratives celebrating our Irish heritage. I was always encouraged to go to medical school, but despite being accepted into pre-med, after taking an introductory module in the social sciences, I was fascinated with the complexity of how health-related issues were being explored in this field. By recognizing the complexity of the social and cultural factors in understanding health-related issues, I agreed that in order to help people, we need to think about the whole person, including their culture and social environments. So I began my interdisciplinary social science health career, culminating in my working with my doctoral advisor, himself a product of the 1950s experiment of eliminating separate social science disciplines, and cross-appointed in anthropology, sociology and psychology, and although my degree says is from the Department of Sociology, my first presentation and publications were at anthropology conferences and in anthropology journals, and the Department of Anthropology at MU housed my Cochrane fellowship and was my home institution for my MSCA, as well as my most recent Fulbright-HRB (Irish Health Research Board) Health Impact scholars award.

As we mentioned at the beginning of this book, our research is based on our MSCA comparative ethnography of donor human milk banking across the UK for the MUIMME project, which was showcased in 2016 for the MSCA 20th anniversary "Science is Wonderful" night at the Parlamentarium in Brussels (which also has a comparatively low rate of breastfeeding). Their chosen theme was to highlight MSCA projects linked to food research, and I made the case that the human milk is food, and they gave me an opportunity to actively engage with the food studies, as well as the health studies communities from across Europe, including talking with 2,000 10-year-olds about the potential lifesaving importance of maternal milk. Gabriel, who was also 10 years old at the time, also spent the entire day helping me talk to his peers in Brussels. The applied personal nature of this aspect of my work is one that I hope to encourage in others, although I am also reminded that the affective and moving components of my experiences are also often gendered and that the role of the mother in the neonatal unit is one which is also often gendered, and therefore even though the human milk she produces is considered invaluable, she, as an affective complication, can still be marginalized and devalued for her role in her own healthcare and that of other infants. Her knowledge and experiences as a mother are different from those of the healthcare community, but they are equally valuable in the rubric of infant health concerns. Medical research is increasingly recognizing the invaluable role of the patient and the public in the expansion of knowledge (Stuttaford et al. 2017), although this is complicated in the case of neonatal care where parental rights are sometimes at odds with medical rules, not to mention the post-traumatic stress that NICU parents may be experiencing (Shaw et al. 2013). These parents, and in particular these mothers, are particularly vulnerable, but they should also be valued. As the value of human milk in the neonatal unit, and for infants in general, expands around the world, we need to not separate this value from the mother-child dyad which is necessary for it to be produced, valuing not only that relationship but also the constituent parts, which needs to be recognized to expand the complexity of our knowledge and understanding of banking on milk.

Tanya Cassidy

Note

1 Full-term birth is estimated to occur after 37 weeks and up to approximately 40 weeks gestation, with preterm birth being defined as occurring before 37 weeks (Martin et al. 2009). Recommendations for age-related corrections, meaning thinking of the infants' age when they were actually born, as opposed to when they were supposed to be born, dates back to the 1930s (Mohr and Bartelme 1930) and continues to be the medically supported vision today (D'Agostino 2010). This is relevant for clinical considerations of developmental milestones.

References

Adhisivam, B., B. Vishnu Bhat, N. Banupriya, Rachel Poorna, Nishad Plakkal, and C. Palanivel. 2017. "Impact of Human Milk Banking on Neonatal Mortality, Necrotizing Enterocolitis, and Exclusive Breastfeeding – Experience from a Tertiary Care Teaching Hospital, South India." *The Journal of Maternal-Fetal & Neonatal Medicine* 32 (6): 902–5.
Anderson, L. 2006. "Analytic Autoethnography." *Journal of Contemporary Ethnography* 35 (4): 373–95.
Cassidy, Tanya, Anne Matthews, and Roslyn Giglia. Forthcoming. "Psychosocial and Cultural Interventions for Reducing Alcohol Consumption during Lactation." *Cochrane Review*.
Cassidy, Tanya and Abdullahi El Tom. 2010. "Comparing Sharing and Banking Milk: Issues of Gift Exchange and Community in the Sudan and Ireland." In Alison Bartlett and Rhonda Shaw (eds.), *Giving Breast Milk: Body Ethics and Contemporary Breastfeeding Practice*. Toronto: Demeter Press, pp. 110–21.
Cassidy, Tanya and Conrad Brunström. 2015. "Production, Process and Parenting: Meanings of Human Milk Donation." In Tanya M. Cassidy and Florence Pasche Guignard (eds.), *What's Cooking Mom? Narratives About Food and Family*. Toronto: Demeter Press, pp. 58–70.
D'Agostino, J. A. 2010. "An Evidentiary Review Regarding the Use of Chronological and Adjusted Age in the Assessment of Preterm Infants." *Journal for Specialists in Pediatric Nursing* 15: 26–32.
Denzin, N. K. 2014. *Interpretive Autoethnography*. 2nd ed. Los Angeles, CA: Sage.

Ellis, Carolyn. 1997. "Evocative Autoethnography." In *Representation and the Text: Re-Framing the Narrative Voice*, edited by W. G. Tierney and Y. Lincolnm, 115–39. Albany: State University of New York Press.

Ellis, Carolyn and Arthur P. Bochner. 1996. *Composing Ethnography: Alternative Forms of Qualitative Writing*, Vol. 1. Walnut Creek: AltaMira.

———. 2000. "Autoethnography, Personal Narrative, Reflexivity: Researcher as Subject." In *The Handbook of Qualitative Research*, edited by N. Denzin and Y. Lincoln. 2nd ed., 733–68. Thousand Oaks, CA: Sage.

Ellis, Carolyn, Tony E. Adams, and Arthur P. Bochner. 2010. "Autoethnography: An Overview [40 Paragraphs]." *Forum Qualitative Sozialforschung/Forum: Qualitative Social Research* 12 (1), Art. 10. http://nbn-resolving.de/urn:nbn:de:0114-fqs1101108.

Martin, J. A., B. E. Hamilton, P. D. Sutton, S. J. Ventura, F. Menacker, S. Kimeyer, et al. 2009. "Births: Final Data for 2006. National Center for Health Statistics." *Vital Health Statistics Series* 57 (7): 1–102.

Mohr, G. J. and P. Bartelme. 1930. "Mental and Physical Development of Children Prematurely Born." *American Journal of Diseases in Childhood* 40: 1000–15.

Reed-Danahay, D. E. ed. 1997. *Auto/Ethnography. Rewriting the Self and the Social.* Oxford: Berg.

Shaw, R., St John, N., E. Lilo, B. Jo, W. Benitz, D. Stevenson and S. Horwtiz. 2013. "Prevention of Traumatic Stress in Mothers with Preterm Infants: A Randomized Controlled Trial." *Pediatrics* 132: e886–e894.

Strathern, M. 1987. "The Limits of Auto-Anthropology." In *Anthropology at Home*, edited by A. Jackson, 16–37. London: Tavistock.

Stuttaford, M. C., T. Boulle, H. J. Haricharan, and Z. Sofayiya. 2017. "Public and Patient Involvement and the Right to Health: Reflections from England." *Frontiers in Sociology*. www.frontiersin.org/article/10.3389/fsoc.2017.00005

Index

Abrams, K. 72
Acharya, C. 93, 98
Adams, T. 116, 123
ADC (Archives of Disease in Childhood) 48
Ades, A.E. 52
Adhikari, M. 113
Adhisivam, B. 5, 45, 88, 120
Africa 18, 59, 104, 115
AIDS 24, 43, 53, 103
Akers, R.M. 64, 69
ALCI (Association of Lactation Consultants of Ireland) 6
AlHreashy, F.A. 107
Allen, L.H. 61, 70
allomaternal 23, 92, 98
Almeida, S.G. 103, 113
AL-Naqeeb, N.A. 51
Altosaar, I. 64, 69
Amundson, K. xi, 19, 70
Anblagan, D. 99
Anderson, L. 116
Andreas, N.J. 61, 64, 69
Appadurai, A. 23
Appelbee, A. 69
Apple, R.D. viii, 23, 42
Arcaro, K. 67
Arck, P.C. 14
Ardythe, L. 69
Arensberg, C. 18–19
Arnold, L. 101–2
Arslanoglu, S. 103, 113
Arthi, V. 26, 51
Asaria, M. 113
Ascroft, E. 32
Ashton, J. 22, 54
Aslan, E. 108, 115
Atkins, P. 56, 69
Atkinson, P. 74, 86
Australian 52, 100, 120
Austria 24, 52

autoethnography 86123
Autran, C. 70
Axelin, A. 98–9

babyfriendly 113, 115
BabyMilkAction 112
Bajaj, R. 5, 22
Balmer, S. 24, 42, 43, 44, 46–7
Banupriya, N. 19, 50
BAPM (British Association of Perinatal Medicine) 4–5, 19, 104–5
Bartelme, P. 122
Bauer, M.W. 99
Beck, U. 16
Becker, G.E. 99
Beiser, F.C. 18–9
Beliu, R. 113
Beltran, E. 71
bereaved 47, 97, 117
bereavement 97–8, 116–17
Bertino, E. 72
Berton, P. 51
bioactive 57, 61, 91
biobank 66, 111
biocapital 3
bioeconomy 3
biopower 3
biovalue 3
Blais, D.R. 69
BMJ (British Medical Journal) 35, 48
Bochner, A. 18, 20, 116, 123
Bode, L. 64, 69
Boston 26, 35, 47, 52
Boswell-Penc, M. 61, 69
Bosworth, A.W. 62
bottles 29, 35–6, 47, 61, 82–5, 93, 94
Boulle, T. 122
Bourgeault, I. xvii
Bourlieu, C. 64, 69
Boyd, C.A. 51

Boyer, K. 61, 69
BPNI (Breastfeeding Promotion Network of India) 114
Bracht, M. 100
Brandstetter, S. 3, 113
Brazil xvi, 15, 103, 113
breast pump 61
Brexit 10, 46, 108–9
Brocklehurst, P. 51
Brown, A. x, 62
Brunström, C. 117
Budin, P. 26
Buffin, R. xi, 57, 100, 114
Butler, J. 14, 22
Butt, M.Z. 115
Butte, N. 51

Cabana, E. 60
Cadwell, K. 101
Cahn, Sir J. 27, 49
Calder, R. 27, 50
Campbell, C.T and T.M. 67
Cannon, A. 64
Capuco, A.V. 64, 69
Carroll, K. 68, 75, 86, 90
Cassidy, T.M. i–iv, vii–ix, xvi–xvii, 3–4, 12, 19, 24, 26, 37, 40, 44, 51, 88–9, 94, 97, 99–100, 106, 107, 113, 117, 120, 122
Castells, M. 15
Chakraborty, S. 106
champions 102–3, 112
Chaplin, C. 12
charity 47, 49–50
Chester 10, 24, 45–7, 89, 93, 103, 105, 107
Choi, S.A. 64, 69
Clarke, J. 40, 52, 61
Clarke, M. 113, 115
Clinton, Rodham H. 18
CMV (cytomegalovirus) 57–8, 91
Cockburn, F. 43
Coffey, A. 86
Cohen, R.S. 4, 46, 52
Cole, T.J. 53
Colebrook, L. 35
colostrum 94, 96
Connell, L. 101
Cossey, V. 57, 69
Coutsoudis, A. 70, 103, 113–14
Cowen-Fletcher, J. 18
Cregan, M.D. 64, 69
Croke, P.V. 114
Crossland, N. 61

Crossley, W. 35
Crowther, S.M. 41–2, 52
cytomegalovirus (CMV) 57–8, 91

Dadhich, J. 114
Dare, E. 10, 32, 35, 49, 91
Daud, N. 3, 19, 107
Davis, M.W. 64, 70
Davis, M.F. 71
Deegan, M.J. 86
Degenhardt, M. 4
Delamont, S. 86
DeMarchis, A. 3, 19, 45, 88, 99, 113
Dempsey, E. 94, 99, 120
Denzin, N.K. 116, 122
Dewey, K. 99
Dionne quintuplets 34, 53
Diprose, R. xvii, 3, 14, 19
Doehler, K. 3
Dowling, S. 106, 113
Dunn, T.D. 43
Dykes, F.C. i–iv, ix, xv, 4–5, 12, 19, 41, 88, 94, 97, 99, 102, 113, 117
Dynski-Klein, M. 48–9, 52

EBM (expressed breastmilk) 79, 93
EL-Khuffash, A. 108, 114
Ellis, C. 20, 116, 123
El Tom xvi–xvii, 51, 86, 99, 107, 117, 120
EMBA (European Milk Banking Association) xi, 24, 56–7, 64
Escherich, T. 25–6, 57, 72
Escuder-Vieco, D. 55, 59, 70
Espinosa-Martos M. 70
Esquerra-Zwiers, A. 21, 100
Eynesbury (St Neot's quadruplets) 33, 50

Fabie, J.E. 71
Faircloth, C. 20
Falls, S. 3
Fantus, B. 47
Feinberg, M. 420
Fields, D.A. 64, 70
Fildes, V. 23, 52
Flacking, R. 98–9
Foster, K. 51
France (lactarium) xi, 18, 20, 5420, 56–7, 60, 71, 91, 116
Franklin, S. 3, 14, 20, 22
Frantz, A.L. 72
Frauenmilch 25, 53, 71
Fuller, S. 5
Fumagalli, S.M. 100
Furman, L.M. 4

Gadner, H. 53
Gates, Bill and Melinda Foundation 21, 112
Geddes, D.T. 2, 64, 69, 71–2
Giglia, R. 106
Glasgow x, 10, 42–5, 105, 108, 110
Goddard, M.K. 105, 114
Golden, J.L. 23, 26
Goldman, A.S. 64, 70
Greenwood Wilson, J. 49
Grishchenko, V. 64, 71
Groer, M. 64
Grønn, M. 65, 70, 77, 86, 107, 114
Grøvslien, A. xi, 43, 65, 70, 77, 114
Gupta, A. 107, 114
Gustafsson, L. 64, 70
Guyer, J. 12, 100

Håkansson 64, 67, 70
Hall, R. 54
Hallett, T. 86
Hallgren, O. 70
Hallman, N. 24
Hamilton, B.E. 122
Hansen, B.M. 100
Haricharan, H.J. 122
Harrison, E.H. 33–5
Hartmann, B.T. xi, 52, 69, 72
Hassiotou, F.A. 2, 64, 71
HAU, Journal of Ethnographic Theory 14, 22
Haussman, B. 5, 22
Helleparth, M. 50
Henderson, G. 21
Hepworth, A.R. 64, 69, 72
Hewlett, B.S. 23, 99
Hill, D.R. 64, 71
Hinde, K. 61
Hirsch, E. 5
HIV (Human Immunodeficiency Virus) 24, 43, 50, 53, 58, 60, 92, 103
HMBANA (Human Milk Banking Association of North America) 67, 101–2
Hobbs, A.J. 92, 99
Holder pasteurization (HoP) 59, 64–5, 81, 72, 81
Holt, L.E. 42, 53
HTLV (human T-cell lymphotropic virus, or human T-cell leukemia-lymphoma virus) 58, 60
HTST (high temperature short time) 59, 65
Hurst, N.M. 22

IBCLC (International Board Certified Lactation Consultant) 104
IBFAN (International Baby Food Action Network) 114
ICCBBA (International Council for Commonality in Blood Banking Automation) 60
Insel, R.A. 64, 71
Iphofen, R. 8
ISBT 128 (International Society of Blood Transfusion) 60
Israel-Ballard, K. xi, 19, 52, 70, 92, 99, 113, 115
iThemba Lethu (I have a destiny) 103–4

Jackson, W. 106, 123
Jacobs, J. 21
Jeffers, G. 74, 86
Jellett, H. 40, 53
Jensen, T. 69
Jones, F. 24
Jovchelovitch, S. 89

Kaiser-Franz-Josef Spital hospital 25
Kakulas, F. 69, 72
Kaminska-E-Hassan, E. 11
Kampmann, B. 61, 69
Kantorowska, A. 420
Kataev, A. 64, 71
Kayser, M.E. 26, 50, 54
Keil, A.D. 52
Kepler, P. 25–6, 53
Kimball, S.T. 18–19
kinship xvi, 18, 60, 107, 119
Kiriline, L. 53
Kleinman, A. xiv, xvii
Koch, J.T. 40
Kohl, S. 64, 72
Kolling, H. 54
Kristeva, J. 13–14, 18
Kronborg, H. 100

Labbok, M. 5, 22
La Leche League (LLL) 5
Lamphere, L. 1421
Latour, B. 18, 56, 71
Lau, C. 22, 94
Lee, R. 20, 64, 69, 106–7, 114
Lepore, J. 61
Lincoln, Y.S. 123
Little, H.M. 37, 52
Litwin, S.D. 64
Lock, M. 3, 20

Lönnerdal, B. 64, 71
Lopez, M. 51, 115
Lucas, A. 50, 53

MacLean, A.M. 86
Mahon, B.P. ix, 7, 113
Malaysia 107
Malindine, E.G. 27, 50
Malinowski, B. 13
Mannion, C.A. 99
Marx, K. 11, 23
Maternowska, B.F. 99
matrescence 14, 98
matricentric 89
Mauss, M. xiv, 12–14, 88, 100
Mayerhofer, E. 25, 53, 57, 71
McAllister, M. 20
McCrea, A. x, 44–5, 53
McDonald, S.W. 99
McGrath, J.M. 69
McGuire, W. 20–1
Mead, M. 101
Mehring, K. 61, 69
Meier, P.P. 4, 20, 88, 100
Merleau-Ponty, M. xvii, 19

Metlinger, B. 37
Michalski, M.C. 64, 69
Miletin, J. 94, 99
Miller, L. 20
Mills, A.J. 8
Miralles, O. 64
Mitanchez, D. 100
Mitchell, R. 3, 22
Mohinuddin, S.M. 115
Mohr, G.J. 123
Moro, G.E. xi, 71, 113
Morrow, A.L. 61, 69
Mosca, F. 61, 71
Mossberg, A.K. 70
Mosse, B. 40
Mota, T. 52
MUIMME (Milk Banking and the Uncertain Interaction between Maternal Milk and Ethanol) xii, xv, 1, 7, 9, 40, 50, 76, 104, 121
Murdoch 106, 114
Murphy, L. 67, 114

Naicker, M.A. 114
Nair, N. 113
National Birthday Trust Fund 23, 26–7, 32

NEC (necrotizing enterocolitis) xii, 4, 11, 93, 118, 120
Newburg, D.S. 64, 71
Newell, M.L. 52
Nordin, N. 19, 114
Novalis (pseudonym for Georg Philipp Friedrich Freiherr von Hardenberg 1772–1801) 18

Oakley, A. 8, 22, 24, 54
O'Brien, L. 98, 100
O'Connor, D.L. 21
O'Connor, R. 91
Oftedal, O.T. 64, 71
O'Reilly, A. 40, 89, 100
Orrenius, S. 64, 70
O'Sullivan, E. 89

Palmer, G. 102
Palmquist, A. 3, 75, 86
Palou, A. 71
Pang, W.W. 52
Park, E.Y. iv, 10, 69
Patel, A.L. 4, 21, 100
Paterson, D. 35, 37
PATH (Program for Appropriate Technology in Health) xi, 3, 21, 100102, 106
Pavone, V. 3
Pehar, A. 69
Percy, C. 53
Picaud, J.C. xi, 57, 100, 114
Porter, R. 36, 53
Přibram, E. 25–6, 53, 57, 71
Propper, C. 113
Provenzi, L.B. 100

Quigley, M.A. 51, 99

Rabinow, P. xv, xvii, 3, 721
Rapp, R. 7, 1421
reciprocity 2, 88–9, 101
Reed-Danahay, D. 116, 123
Reimers, P. 103, 115
Reynolds, L.A. 41–2, 52
Robinson, R. 43
Robson, K. 100
Rodham Clinton, H. 18
Rodríguez, J.M. 70
Rogier, E.W. 64, 72
Rosack, M.L. 53
Rosaldo, Z.M. 14
Rose, N. 321
Rossi, L. 72

Rotch, T.M. 6, 21
Rotunda hospital, Dublin Ireland 40, 53
Rubin, G. 14

Sacks, A. 14
Sanchez, J. 71
Sanlaville, D. 114
Sardo, A. 72
Savino, F. 72
Schanler, R.J. 72
Schmidt, S. 70
Schneider, E. 26, 51
Schon, D.A. 102
Schuman, A. 54
Scotland x, 1, 12, 24, 42–6, 60, 103, 105, 108
Seifert, R. 26, 54
Senol, D.K. 108, 115
Shaw, R. xvii, 3, 18, 22, 68, 102, 123
Shenker, N.S. 65, 72, 115
Shulman, S.T. 57
Silverman, W.A. 26
Sims, R.H. 72
Slimes, M.A. 24
Smith, G.D. 53, 69
Smith, H.A. 99
Smith, J.F. 37
Smith, L.B. 114–15
Smith, P.H. 5
Snyder, J.R. 6, 22
Solomons, B. 40
Springer, S. 19, 87, 113
Stelzer, I.A. 14
Strathearn, L. 72
Strathern, M. xiv, xvii, 14, 22, 116, 123
Strong, T. 22
Sullivan, G. 99
Sussman, G. 23, 54
Sutton, P.D. 123
Svanborg, C. 64, 67, 70
Swanson, K.W. xv, xvii, 10, 23, 47

Tansey, E.M. 41–2, 52
Tchaikowski, C. 63
Thomson, G. 8, 98–9
Thorley, V. 61, 107, 115
Titmuss, R. 8, 13, 22–4
Tobey, J.A. 23
trust xvi, 1, 12, 15, 18, 67–8
Twigger, A.J. 64

UKAMB (United Kingdom Association for Milk Banking) xi, xiii, 9, 24, 44, 46–7, 50, 67, 89, 108, 109, 117
Unger, S. 108, 114
UNICEF vii, 42, 62, 103–4, 106–7
Unvas-Moberg, K. 64, 72
Utrera Torres, M.I. 115

Van Esterik, P. 61, 91
Vanhole, C. 69
Van Slyke, L.L. 62
Vargas-Martínez, F. 64, 72
Verhasselt, V. 64, 72
Vienna 24–6, 50–1, 57
Vietnam 105–6
Vogel, H.J. 72

Waldby, C. 3, 22
Ward, J. 5, 22
Watson, M.L. 47, 54
WBTi (World Breastfeeding Trends initiative) 102, 114
Weaver, G. x, 8, 22–3, 27, 37, 50, 103–4, 115
Webster, C. 49
Wei, J.C. 4
Wesolowska, A. xi, 109
wet nursing 23–4, 37, 40, 54, 120
Wharton, B.A. 24
WHO (World Health Organization) xiii, 23, 42, 54, 60, 107
Wight, N.E. 54
Williams, A.S. 8, 23, 37, 50
Williams, K. 77, 86, 87–8
Williams, T.C. 107, 115
Wilson-Clay, B. 54
Winberg, J. 64, 72
Winn, S. 23, 52, 99
Wright, K. 4

Yusoff, Z.M. 19, 114

Zehr, B.D. 64, 71
Zherelova, O. 64, 71
Zhivotovsky, B. 70
Ziegler, J.B. 43
Zipitis, C.S. 5, 22
Zizzo, G. 94